INQUIRING

SCIENTISTS,

INQUIRING

READERS

Using Nonfiction to Promote Science Literacy

GRADES 3–5

Fries-Gaither & Shiverdecker

INQUIRING
SCIENTISTS,
INQUIRING
READERS

Using Nonfiction to Promote Science Literacy
GRADES 3–5

Jessica Fries-Gaither
Terry Shiverdecker

National Science Teachers Association
Arlington, Virginia

National Science Teachers Association

Claire Reinburg, Director
Jennifer Horak, Managing Editor
Andrew Cooke, Senior Editor
Wendy Rubin, Associate Editor
Agnes Bannigan, Associate Editor
Amy America, Book Acquisitions Coordinator

ART AND DESIGN
Will Thomas Jr., Director
Rashad Muhammad, Graphic Designer, cover and interior design

PRINTING AND PRODUCTION
Catherine Lorrain, Director
Nguyet Tran, Assistant Production Manager

NATIONAL SCIENCE TEACHERS ASSOCIATION
Gerald F. Wheeler, Executive Director
David Beacom, Publisher
1840 Wilson Blvd., Arlington, VA 22201
www.nsta.org/store
For customer service inquiries, please call 800-277-5300.

NSTA is committed to publishing material that promotes the best in inquiry-based science education. However, conditions of actual use may vary, and the safety procedures and practices described in this book are intended to serve only as a guide. Additional precautionary measures may be required. NSTA and the authors do not warrant or represent that the procedures and practices in this book meet any safety code or standard of federal, state, or local regulations. NSTA and the authors disclaim any liability for personal injury or damage to property arising out of or relating to the use of this book, including any of the recommendations, instructions, or materials contained therein.

Library of Congress Cataloging-in-Publication Data

Shiverdecker, Terry, 1956-
  Inquiring scientists, inquiring readers : using nonfiction to promote science literacy, grades 3-5 / by Terry Shiverdecker and Jessica Fries-Gaither.
     pages cm
  Includes bibliographical references and index..
  ISBN 978-1-936959-10-5
  1. Scientific literature--Study and teaching (Elementary) 2. Science--Study and teaching (Elementary) I. Fries-Gaither, Jessica, 1977- II. Title.
  Q225.5.S55 2012
  372.35--dc23
                              2012025621
eISBN 978-1-936959-54-9

CONTENTS

CONTENTS

ACKNOWLEDGMENTS

We would like to thank the following individuals for their support and assistance:

- Our husbands, Tre' Gaither and Mark Shiverdecker, and families for their love, support, and infinite patience while we disappeared under a pile of books and materials.

- Kimberly Lightle and Nicole Luthy for their support and encouragement.

- Judy Duguid and Carolyn Hamilton for their editing expertise.

- Ohio Resource Center staff for their support, encouragement, and good humor.

- Faye Harp for her extensive review of the inquiry units.

- Susan Putnam, Shannon Patterson, Margie Belair, Russell Long, Mary Rice, Patricia Vasquez, Julie Robinson, Carrie Waugh, and Jamie Dean for field-testing the inquiry units.

- Meri Johnson, Ricky Austin, Meghann Robinson, Elizabeth Fries, Michele Weber, Dee Martindale, Cheryl Zachry, and Jane Galbraith for their assistance in recruiting field test teachers.

ABOUT THE AUTHORS

Jessica Fries-Gaither is the Lower School science teacher at Columbus School for Girls (CSG) in Columbus, Ohio, where she teaches science to grades 1–5. Prior to working at CSG, she was an education resource specialist in the School of Teaching of Learning, College of Education and Human Ecology at Ohio State University, where she served as the project director for two National Science Foundation–funded projects: *Beyond Penguins and Polar Bears* and *Beyond Weather and the Water Cycle*. Both projects support elementary teachers as they integrate science and literacy instruction. Jessica earned bachelor's degrees in the biological sciences and anthropology from the University of Notre Dame. She received her MEd from the University of Notre Dame's Alliance for Catholic Education program. Jessica taught middle school math and science in Memphis, Tennessee, and a variety of grade levels, ranging from eighth-grade science to fourth-grade self-contained, in Anchorage, Alaska.

Terry Shiverdecker is the science content specialist at the Ohio Resource Center located at Ohio State University. She provides teachers of science in grades K–12 best-practices resources and professional development opportunities. Terry received a bachelor's degree from Wright State University, a master's degree from Miami University (Ohio), and an EdD in curriculum and instruction from the University of Cincinnati. She taught high school science for 12 years at Russia Local School in west central Ohio and has experience teaching both physical and life sciences. Terry also worked at the Shelby County (Ohio) Educational Service Center as Director of Secondary Curriculum and Instruction.

INTRODUCTION

Every now and then you happen on an idea that ties multiple lines of thought together into a unified whole. An idea that paints a clear picture, answers questions, and leads you in a new direction. We encountered such an idea when we read the article "Science Text Sets: Using Various Genres to Promote Literacy and Inquiry," by Margaretha Ebbers (2002). Ebbers describes how multiple genres of nonfiction (including biographies, field guides, reference books, and journals) present scientific information in different ways. Collections of these books, known as text sets, can be used to support inquiry-based instruction by assisting students as they pose questions, design investigations, and confirm and extend the knowledge they've learned through direct investigation.

This article transformed our thinking and got us excited about the possibilities for elementary science education. Jessica, an elementary school teacher and the project director of two elementary science and literacy projects, was thrilled to discover a new way to promote the use of nonfiction and informational texts in the elementary grades. Terry, a science content specialist and former high school science teacher, was pleased to find a way to integrate science and literacy instruction without sacrificing the science. As we planned a three-day workshop on the subject for elementary teachers, we became increasingly convinced that this approach held great promise. We received very positive feedback from the workshop and from several additional presentations at local conferences. We knew that we wanted to share this approach with as many teachers as possible—and the idea for this book was born.

ABOUT THIS BOOK

In this book we will weave together best practices in literacy and science instruction in a way that makes sense for classroom practice. Inquiry is at the heart of best practices in science instruction. The units in this book are inquiry-based units, involving testable questions and the use of evidence to construct conclusions. In most cases, the teacher provides students with a question and a procedure. In some cases, the teacher supports students as they pose questions, design an investigation, and collect evidence. We designed teacher-led inquiries in an effort to support teachers new to inquiry. In Chapter 2 we have provided general suggestions for moving the units toward being student directed.

Our inquiry units have been designed using the learning cycle framework, a process that incorporates hands-on investigations, reading science text, directed discussion, and problem solving. The hands-on investigations and reading of science texts are seamlessly integrated, supporting both science and literacy. Sometimes the inquiry units start with a hands-on investigation, other times they start with an engaging nonfiction text. Either way, they follow the learning cycle framework. In terms of literacy, the units emphasize the use of nonfiction text, a genre that has traditionally been underrepresented in the elementary grades. They also provide authentic contexts for reading, writing, and discussion through read-alouds, collaborative activities, graphic organizers, and writing prompts. Uniting the best practices in science and literacy instruction elevates learning in both areas.

We begin with a review of the literature on science and literacy instruction. We look at inquiry-based instruction and the history and application of the learning cycle. We review some best practices in literacy and how these practices manifest themselves in the units. The foundational chapters also include information about nonfiction genres and text sets.

This research is then applied to the units that follow. This is where the theory is turned into practice. We know that for many teachers this will be the most enticing part of the book. We have included units that encompass life, physical, and Earth and space science as well as two on the nature of science. Each inquiry unit includes scientific background information, common misconceptions that are associated with the content, an annotated list of the texts in the text set, safety considerations, supporting documents (e.g., graphic organizers, writing prompts), and suggested assessments. The units have been tested in classrooms to ensure their effectiveness.

WHY THIS BOOK IS IMPORTANT

As former teachers and current professional development providers, we know that the elementary curriculum has become increasingly crowded over the past few years. An intense focus on literacy and mathematics, due in part to mandated standardized testing, has left little time for science instruction. Much to our dismay, studies indicate that only a limited amount of science is being taught in the elementary grades. From the perspective of science educators, this is problematic. We need to have good science instruction in these formative years for some of the same reasons we need good literacy and mathematics instruction early in a child's formal education. One potential remedy to the problem is to integrate science into literacy instruction. It seems like a good idea, but this approach often leads to reading about science rather than engaging in scientific inquiry. Another potential pitfall with this approach is that the children's literature selected for science reading may be scientifically inaccurate or fiction. While fiction is engaging and plays an important role in early literacy instruction, we believe that scientific content and the nature of science are best conveyed through nonfiction and informational genres.

Science is an active process that must be experienced to be fully understood. Just reading about science is inadequate and cannot be substituted for effective science instruction. Yet it is clear that children must be able to make meaning from text and visual representations of scientific ideas; they must be able to write to express their understanding of science; and they must have the skills to engage in discourse around scientific ideas and processes. It is impossible to help a child develop a strong foundation in science content and processes without drawing on these literacy skills. Science education has struggled with how to build science and literacy skills concurrently in a cohesive manner that honors the best practices of both disciplines.

Teaching nonfiction text comes with its own set of challenges. These challenges include available instructional time, identification of appropriate reading materials, and providing authentic contexts within which the reading can occur. It could be argued that some sci-

INTRODUCTION

ence time could be "borrowed" for instruction in nonfiction text. The process we propose in this book does not require that time be borrowed from either discipline to support the other. This approach supports both equally well. It also reflects what scientists do. Reading and writing is an integral part of science. The issue of identifying appropriate reading materials tends to take care of itself once the topic has been selected. The caveat here is that the reading material must present science content accurately. The issue of providing authentic contexts for reading and writing is also resolved with the careful selection of inquiry topics and tasks. This approach is a win-win for both disciplines.

HOW TO USE THIS BOOK

This book is composed of two distinct sections: the research section and the inquiry units. In a perfect world, the reader would begin by reviewing the research background we've provided and follow that by trying some of the units. But we know what we would do with this book if we were still in the classroom: We'd start with the units. And we know from our research into reading nonfiction and informational text that this makes perfect sense. After all, if you start with a lesson, you will then have an authentic context in which you can place the research. Research always makes more sense when you can put it into a meaningful context.

Even if you choose to start with the units, please go back and read the research review we've provided. Digging into the lessons may be more exciting but your understanding of why the lessons are laid out as they are, why specific practices are suggested, and why certain texts have been selected will be enhanced if you become familiar with the research. The research foundation will also be helpful if you find you need to modify the units to better meet the needs of your students.

Embedding multigenre nonfiction text sets into inquiry instruction is the logical next step toward uniting the two into a cohesive whole. This approach honors the best practices of both disciplines, provides an authentic context for literacy instruction, and supports inquiry into concepts that cannot be easily investigated directly. It is our hope that you will feel empowered and impassioned after reading this book, that you will feel compelled to try all of the lessons that fit your curriculum, and that you will be inspired to develop your own multigenre nonfiction text sets and associated inquiry investigations.

We began this introduction by talking about how the Ebbers article painted a clear picture, answered questions, and led us in a new direction. We hope this book does the same for you.

REFERENCE

Ebbers, M. 2002. Science text sets: Using various genres to promote literacy and inquiry. *Language Arts* 80 (1): 40–50.

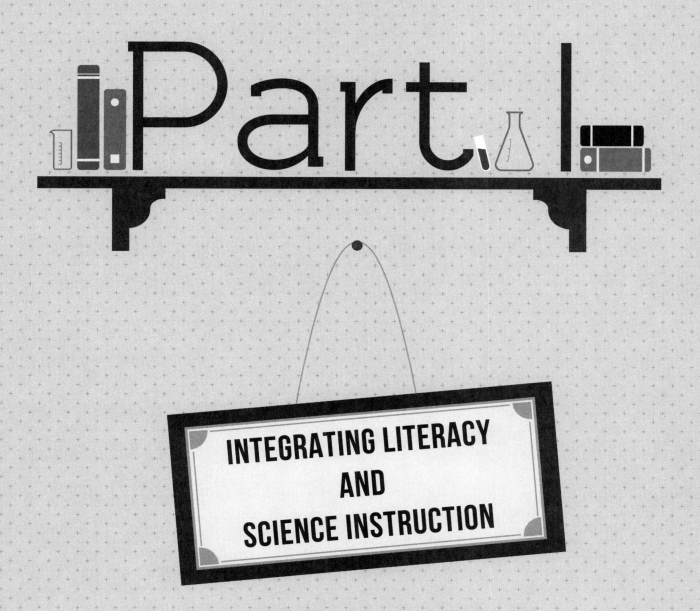

Part I

INTEGRATING LITERACY AND SCIENCE INSTRUCTION

Chapter 1
The Science-Literacy Connection

There has been much talk about integrating science and literacy instruction in recent years. Why this focus, particularly in the elementary classroom? One very practical reason is to make time in a crowded curriculum. Teachers are confronted with the challenge of having so much content to teach that there's simply not enough time in the day. An increased emphasis on reading and the English language arts (ELA) in the form of 90- to 120-minute literacy blocks and 60-minute mathematics classes frequently results in science and social studies getting short shrift. Research shows that the amount of instructional time devoted to science in kindergarten through grade 5 has declined significantly in the last two decades.

In 1993–1994, 50% of core instructional time was allocated to reading/ELA, 24% to mathematics, and 13% each to science and the social studies (Perie, Baker, and Bobbitt 1997, p. 8). From 2001–2002 through 2006–2007, reading/ELA instruction changed from 378 to 520 minutes per week, increasing to 47% of the total instructional time; mathematics instruction changed from 264 to 352 minutes per week, increasing to 33% of the total instructional time; and science instruction changed from 226 to 152 minutes per week, a 43% *decrease* in the total instructional time (Jones and Swanson 2009, p. 165). In 2007 elementary schools in the Bay Area reported that 80% of K–5 teachers spent 60 minutes or less on science each week; 16% taught no science at all (Dorph et al. 2007, p. 1). In a study of 10 teachers who were highly engaged in science professional development, Jones and Swanson (2009, pp. 178–179) found that on average the teachers spent 1.64 hours per week on science instruction. Over the course of the academic year no science was taught on 45.7% of teaching days.

One way to address this inequity is through thoughtful integration. Integrating the two disciplines ensures that science receives sufficient time while still meeting the requirements of the literacy curriculum. There are also a number of sound pedagogical reasons to integrate the two disciplines. First, there is a natural overlap between the two in many areas. Consider inquiry, a way of teaching and learning that lies at the heart of science. The *National Science Education Standards* (NRC 1996, p. 23) provides this explanation of inquiry:

> Inquiry is a multifaceted activity that involves making observations; posing questions; examining books and other sources of information to see what is already known; planning investigations; reviewing what is already known in light of experimental evidence; using tools to gather, analyze, and interpret data; proposing answers, explanations, and predictions; and communicating the results.

Compare that with the explanation of nonfiction inquiry presented in Stephanie Harvey's *Nonfiction Matters: Reading, Writing, and Research in Grades 3–8* (1998, p. 2):

> Inquiry requires that...students explore their passions and curiosity, pursue topics of interest, ask questions, conduct primary and secondary research, read for meaningful content, organize and synthesize information, craft authentic reports, and present and report findings—and gain new understanding in the process.

While the details of how students learn about a topic differ (experiments and investigations in science, research in ELA), it is clear that the two have much in common. In fact, most of Harvey's description of nonfiction inquiry is included in the explanation of scientific inquiry from the *National Science Education Standards*. A well-structured inquiry experience can meet both of these descriptions and support learners as they develop important skills and strategies.

There are several skills and strategies that are fundamental in both scientific and nonfiction inquiry. Padilla (1990) lists observing, inferring, measuring, communicating, classifying, and predicting as basic science process skills. (See Appendix 1 [p. 269] for more information about science process skills.) Several of these (i.e., inferring, communicating, classifying, and predicting) have a place within reading and ELA instruction. These skills may take on a slightly different flavor in each discipline, but they still represent the same thinking skills, regardless of the context in which they are applied (Ostlund 1998). As educators, we want to facilitate students' abilities to apply skills and knowledge from one context to another. Integrating science and literacy may encourage transfer of these critical-thinking skills. As Ostlund (1998) notes, "When a teacher helps students develop scientific processes, reading processes are simultaneously being developed."

Scientific inquiry provides an authentic context for reading, writing, and dialogue. Having a compelling reason to read a book, record observations, and communicate with others increases students' motivation to engage in nonfiction reading. When teachers provide students with authentic reading and writing tasks, students experience growth in reading comprehension and writing skills (Purcell-Gates, Duke, and Martineau 2007).

Although there is substantial overlap between the two disciplines, there are also key differences to which attention must be paid. One such difference is the use of language. Scientific language differs from the language that students use in everyday life (Gee 2004; Varelas et al. 2008) and in language arts instruction. Language in science must be unbiased and precise. Scientists talk about the natural world using quantitative data (e.g., the flower has five petals) and specific qualitative descriptors (e.g., the petals are a deep blue) instead of words such as *beautiful* or *small*. Additionally, many words have different meanings in science than they do in common language. Consider, for example, the word *theory*. In science, a theory is an explanation supported by a considerable body of evidence. Scientific theories are so well supported that new evidence is not likely to change them substantially. Atomic theory, cell theory, and the theory of plate tectonics are several examples. Contrast this with the way that the word is used in common language: as a supposition that often lacks an evidentiary support. It is especially imperative that teachers wishing to integrate literacy into science understand these distinctions and take care to promote the use of scientific language in their instruction.

Integrating science and literacy supports student achievement in both disciplines. Both reading and scientific inquiry are concerned with thinking processes; both are an effort to construct meaning. Science and reading process skills are developed concurrently in integrated science and nonfiction inquiries (Mechling and Oliver 1983; Ostlund 1998; Simon and Zimmerman 1980). Manipulating materials through scientific inquiry aids in the development of language arts skills (Ostlund 1998; Wellman 1978). Scientific inquiry experiences have a positive effect on vocabulary enrichment, increased verbal fluency, enhanced ability to think logically, and improved concept formation and communication skills (Campbell 1972; Kraft 1961; Olson 1971; Quinn and Kessler 1976; as cited in Ostlund 1998).

Effective integration of science and literacy requires careful consideration of the timing and sequence of instructional activities. One danger of the push to integrate is that the reading of science-themed text replaces the actual *doing* of science. Simply reading about science does not suffice to build content knowledge, develop science process skills, or create an understanding of the nature of science. Furthermore, if the scientific investigation follows reading, the investigation is often no longer an inquiry experience, but an attempt to confirm what was presented in the text. Reading and discussion of text might be used to set the context for or to inform the investigation, but care should be taken to preserve the joy of discovery by avoiding extensive reading before investigations have taken place. Simply put, do first and read later (Yore 2004). Another danger is tacking an inauthentic writing assignment, such as a worksheet or series of low-level questions, on to the end of a hands-on science activity. This approach serves neither science nor literacy. It fails to capitalize on the multitude of opportunities that exist for reading, writing, and conversation around the scientific concepts under study.

In a true integration of science and literacy, the two disciplines are so seamlessly intertwined that separating them would weaken the overall learning experience. Scientific inquiry, including observations and hands-on activities, instill in students a "need to know," an authentic context for reading and writing. Nonfiction and informational texts are found throughout the investigation for a variety of purposes: to engage, to provide structure and guidance for an investigation, and to confirm, clarify, and extend what students have learned from firsthand investigations. Student-directed conversation is a large component of the learning experience. Students talk to the teacher and to each other, comparing findings, questioning, and clarifying. The teacher facilitates many of these discussions, using open-ended questioning to clarify and extend student thinking. The result is no longer "science instruction" or "literacy instruction," but simply effective instruction.

REFERENCES

Dorph, R., D. Goldstein, S. Lee, K. Lepori, S. Schneider, and S. Venkatesan. 2007. *The status of science education in the Bay Area: Research brief*. Berkeley: Lawrence Hall of Science, University of California, Berkeley. *www.lawrencehallofscience.org/rea/bayareastudy/pdf/final_ to_print_research_brief.pdf*

Gee, J. P. 2004. Language in the science classroom: Academic social languages as the heart of school-based literacy. In *Crossing borders in literacy and science instruction,* ed. E. W. Saul, 13–32. Newark, DE: International Reading Association.

Harvey, S. 1998. *Nonfiction matters: Reading, writing, and research in grades 3–8.* Portland, ME: Stenhouse.

Jones, R., and E. Swanson. 2009. Understanding elementary teachers' use of science teaching time: Lessons from the Big Sky Science Partnership. *The Journal of Mathematics and Science: Collaborative Explorations* 11: 163–192.

Mechling, K. R., and D. L. Oliver. 1983. *Handbook I: Science teaches basic skills.* Washington, DC: National Science Teachers Association.

National Research Council (NRC). 1996. *National science education standards.* Washington, DC: National Academies Press.

Ostlund, K. 1998. What the research says about science process skills. *Electronic Journal of Science Education* 2 (4). *http://ejse.southwestern.edu/article/view/7589/5356*

Padilla, M. J. 1990. The science process skills. *Research Matters—to the Science Teacher,* No. 9004. *www.narst.org/publications/research/skill.cfm*

Perie, M., D. Baker, and S. A. Bobbitt. 1997. *Time spent teaching core academic subjects in elementary schools: Comparisons across community, school, teacher, and student characteristics.* Washington, DC: U.S. Dept. of Education, Office of Educational Research and Improvement.

Purcell-Gates, V., N. K. Duke, and J. A. Martineau. 2007. Learning to read and write genre-specific text: Roles of authentic experience and explicit teaching. *Reading Research Quarterly* 42 (1): 8–45.

Simon, M. S., and J. M. Zimmerman. 1980. Science and writing. *Science and Children* 18 (3): 7–9.

Varelas, M., C. C. Pappas, J. M. Kane, A. Arsenault, J. Hankes, and B. M. Cowan. 2008. Urban primary-grade children think and talk science: Curricular and instructional practices that nurture participation and argumentation. *Science Education* 92 (1): 65–95.

Wellman, R. T. 1978. Science: A basic for language and reading development. In *What research says to the science teacher*, Vol. 1, ed. M. B. Rowe. Washington, DC: National Science Teachers Association.

Yore, L. 2004. Why do future scientists need to study the English language arts? In *Crossing borders in literacy and science instruction,* ed. E. W. Saul, 71–94. Newark, DE: International Reading Association.

Chapter 2
Inquiry and the Learning Cycle

One evening when I was in the fifth grade we had roast chicken for dinner. My best friend and I decided that we wanted to reconstruct the chicken skeleton. We cleaned the bones and worked and worked to make them fit together. We were not able to reconstruct the skeleton, but we learned a great deal about the features of bones. My friend is now a nurse, and I, of course, am in science education.

Several years ago, when he was three years old, my grandson and I were putting together a holiday decoration. The decoration included a hollow ball. He picked up the ball, shook it, and heard a rattle. He quickly said, "Grandma, there's something inside." He had just made an evidence-based conclusion.

When my granddaughter was eight years old, she was fascinated with toads. One summer day, she came to me with a toad in each hand. She said, "Listen, Grandma, they make different sounds." Sure enough, the toads' trills were different. She then said, "I think since they make different sounds, this one must be a boy and this one is a girl." They were actually two different yet very similar species. But through careful and patient observation, she noticed differences and drew conclusions based on her observations.

Children conduct inquiries all the time. They make evidence-based decisions daily. Our task is not to teach them how to do inquiry. Our task is to nurture their natural talents.

—Terry

THE FIVE FEATURES OF INQUIRY

Inquiry instruction combines the teaching of scientific content with the development of scientific practices. Inquiry-based instruction has five essential features that occur along an inquiry continuum. On one end of the continuum the essential features are teacher directed, on the other end they are student directed (Center for Science, Mathematics, and Engineering Education 2000). Below we list the five features of inquiry, along with descriptions of the teacher-student relationships for each end of the continuum.

1. *Learners are engaged by scientifically oriented questions.* Scientifically oriented questions are often referred to as testable questions. They are questions that students can investigate through either firsthand or secondhand investigations. Firsthand investigations are those in which students interact directly with the phenomena, making observations, taking measurements, and gathering data. Secondhand investigations are text based (Hapgood, Magnusson, and Palincsar 2004). On the teacher-directed side of the continuum, the question is provided for the student, either by the teacher or the

instructional materials (e.g., science kit, activity worksheet). On the student-directed side, the student constructs the question.

2. *Learners give priority to evidence when developing and evaluating explanations that address scientifically oriented questions.* Students make observations, take measurements and record data, or obtain data from their teacher or another source. On the teacher-directed side of the continuum, the teacher provides the student with the data and instructs them in how to analyze the data. On the student-directed side, the student decides what the evidence will be and how it will be collected. The student analyzes the data independently.

3. *Learners formulate explanations from evidence to address scientifically oriented questions.* Students collect evidence through active investigations. They link this evidence to prior knowledge, constructing new knowledge in the process. Explanations are based on this new knowledge. On the teacher-directed side of the continuum, the student is provided evidence and told how to use the evidence to formulate an explanation. On the student-directed side, the student gathers evidence independently, summarizes the data, and formulates an explanation.

4. *Learners evaluate their explanations in light of alternative explanations, particularly those reflecting scientific understanding.* Students consider alternative explanations offered by classmates, engage in discussions to clarify understanding, and compare their findings and thinking with credible sources. Students then reflect on these explanations and revise their thinking if necessary. On the teacher-directed side of the continuum, the student is given the alternative explanations from credible sources. The teacher explains the connections between the students' explanations and the alternative explanations. On the student-directed side, the student independently seeks out alternative explanations from credible sources. They then make connections between the alternative explanation and their own explanations.

5. *Learners communicate and justify their proposed explanations.* Sharing explanations helps students clarify their thinking, ask and answer content-specific questions, and draw relationships between the evidence, accepted scientific knowledge, and their explanation. On the teacher-directed side of the continuum, the student is given steps and procedures for communicating explanations. On the student-directed side, the student independently forms reasonable and logical arguments to communicate explanations.

In Table 2.1, we use the "Drip Drop Detectives" inquiry unit (Chapter 13) to illustrate how the essential features of inquiry would fall on the continuum.

The five essential features of inquiry can fall anywhere on the inquiry continuum in a single inquiry-based learning cycle lesson or unit. A common myth about inquiry is that it is unstructured and chaotic. In reality, well-planned inquiry-based instruction supports

2

TABLE 2.1. ESSENTIAL FEATURES AND TEACHER- AND STUDENT-DIRECTED ROLES IN "DRIP DROP DETECTIVES"

Essential Features	Teacher-Directed ←——————→ Student-Directed	
1. Scientifically oriented questions	The question is provided for students near the end of the engage phase.	
2. Priority to evidence	On the first day of the explore phase students collect data while conducting investigations that have been planned for them.	On the second day of the explore phase students design investigations and determine what data they will collect and how they will collect it.
3. Formulating explanations from evidence	In the explain phase students write an article that uses evidence collected to answer the question.	
4. Evaluating explanations		Evidence from the second day of the explore phase is compared with evidence from the first day. Conclusions are revisited and confirmed or revised.
5. Communicating and justifying explanations	Near the end of the explore phase students use a variety of texts to confirm and clarify what they have discovered.	

student learning and helps students develop the skills needed to move from teacher- to student-directed learning experiences. The inquiry units in this book are often teacher directed. We wanted to provide adequate support for teachers and students who are new to inquiry. Any inquiry unit in the book can be modified to be more student directed in a variety of ways, as described in the next section. Of course, it's essential to consider students' abilities and prior experiences when making modifications. Move students toward the student-directed end of the inquiry continuum when their skill and comfort level indicates they are ready for a greater degree of independence.

Moving Toward Student-Directed Inquiry

Strategies to make the inquiry units more student directed are provided under each of the five essential features in the following list:

1. *Scientifically oriented questions:* A question is included with each inquiry unit. You can make this feature more student directed by asking students to list questions and then work together to select one that meets the learning objectives.

2. *Priority to evidence:* In most cases in this book we determine what evidence should be collected and provide data collection sheets. You can make this more student directed by discussing with students what evidence could possibly be collected and then decide as a class or by group how to proceed. In lieu of the data collection sheets, you can have students develop procedures, tables, and other documents in science notebooks.

3. *Formulating explanations from evidence:* We have provided strategies and supporting documents for students to use while preparing explanations. You can make this feature more student centered by providing several options for preparing explanations and allowing the students to choose, providing only prompts without supporting graphic organizers, or you can simply ask students to use the evidence they have collected to answer the question being investigated.

4. *Evaluating explanations:* In most cases students are collecting multiple lines of evidence. They look across the evidence to form explanations. They also compare their explanations with text-based information. You can make this more student directed by encouraging students to seek out evidence collected through additional investigations they have designed or from other sources.

5. *Communicating and justifying explanations:* We have provided activities and supporting documents for students to communicate their findings. You can make this more student directed by providing multiple opportunities for students to choose from. You can have students review one another's explanations and ask clarifying questions to increase the level of scientific argumentation. You can expand the audience the students communicate with by inviting scientists and other community members to participate in interactive presentations. You can use web-based publishing tools to share with a broader audience.

Hands-On Activities Versus Inquiry-Based Instruction

It is sometimes difficult to distinguish between hands-on activities and inquiry-based instruction. Hands-on activities may ask students to collect evidence, but students are rarely asked to formulate and communicate evidence-based conclusions. Evidence, explanation, and communication are at the heart of inquiry-based instruction. You know it's inquiry if students are collecting evidence, making evidence-based explanations, and communicating their findings.

THE LEARNING CYCLE

The five essential features of inquiry go hand in hand with the learning cycle framework. The learning cycle has been a staple for instructional planning in science for over 50 years. In a roundabout way, its development is the result of a physicist's visit to his daughter's second-grade classroom. In 1957 Robert Karplus, a physicist from Berkeley, visited his daughter's class to discuss electricity. That visit sparked Karplus's interest in how children learn science. His early experiences in developing and teaching elementary science lessons led him to wonder how learning experiences could be developed that would help a child connect what he or she already knew with accepted scientific knowledge. This wondering led to the development of the learning cycle, a method of guided discovery consisting of three phases: exploration, invention, and discovery (Karplus and Fuller 2002).

The learning cycle has undergone several changes since its introduction. The changes reflect new understandings about teaching and learning. For example, the original learning cycle did not take into account preexisting student ideas or teachers as diagnosticians of student learning (Duschl, Schweingruber, and Shouse 2007). We now know through misconception and formative assessment research that these ideas must be taken into consideration. Later versions of the learning cycle incorporate these ideas.

In the 1980s the Biological Sciences Curriculum Study (BSCS) revised the learning cycle into a 5E approach. The revised version built on the Karplus model by adding phases that access students' prior knowledge and evaluate student learning. The 5E Model consists of engagement, exploration, explanation, elaboration, and evaluation phases (Bybee et al. 2006). The 4E approach from Martin, Sexton, Wagner, and Gerlovich (1997) is comprised of exploration, explanation, expansion, and evaluation phases. In this model the engagement phase is rolled into exploration, and evaluation is an ongoing phase in which the students' conceptual understanding and process skills are assessed. The ongoing evaluation phase of the 4E learning cycle promotes embedded formative assessment. Other versions of the learning cycle include a 7E approach (Eisenkraft 2003) that adds elicit and extend phases and a 6E approach (Chessin and Moore 2004) that integrates technology throughout the learning cycle in an ongoing e-search phase.

OUR APPROACH: ENGAGE, EXPLORE, EXPLAIN, EXPAND, ASSESS

Our approach closely resembles the BSCS 5E Model, but is informed by Martin's approach to embed assessment throughout. We view assessment as central to the learning cycle as opposed to being a final evaluative phase. Whereas assessment is the ongoing gathering of data that are used to inform teaching and learning, evaluation is a judgment of student performance, often resulting in a grade. With the emergence of formative assessment and its positive impact on learning, we feel assessment must be ongoing and embedded. Figure 2.1 (p. 16) illustrates our approach, and an explanation of each phase is provided in this section.

FIGURE 2.1. OUR APPROACH TO THE LEARNING CYCLE

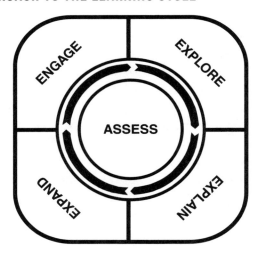

Adapted from the Ohio Resource Center for Mathematics, Science, and Reading. 2012. The Learning Cycle. *www.ohiorc.org/pm/science/Sci_LearningCycle.aspx*

- **Engage:** Activities in the engage phase are designed to access prior knowledge and expose misconceptions; build interest, excitement, and enthusiasm for the topic; and generate a need to know more. Activities could include demonstrations; discrepant events; a read-aloud of an interesting text, poem, or current news story; games; or other activities that introduce the topic and begin conversation. We have included experiences such as read-alouds, videos, a mystery letter, making ice cream, drawings, and scavenger hunts in the engage phase of our inquiry units.

- **Explore:** The explore phase is a time of active investigation. Students are working to answer a testable question. Science process skills are nurtured as students make observations, collect data, develop graphs, make predictions, and draw conclusions through first- and/or secondhand investigations. Students are developing conceptual knowledge in this phase. Scientific terminology is introduced in a just-in-time fashion so that the terms are provided when the students are ready for them. Students now have a meaningful context in which to use scientific language. They will not have mastered the terms at this time. Measurable progress toward achieving the learning objectives becomes apparent in the next phase. In our inquiry units, students engage in a wide range of experiences that build science process skills while revealing the concepts being studied.

- **Explain:** In the explain phase students make evidence-based claims as they share what they have learned using scientifically accurate terminology. The claims are based on the data collected in the explore phase. This phase often involves the development of a product that provides evidence of learning. Some of the products students develop in the explain phase of our inquiry units include responses to writing prompts, explanatory reports, presentations, a class guide to trees, cause-and-effect statements, interviews for a news report, and an article.

- **Expand:** The expand phase provides opportunities to situate new learning in real-world contexts, apply knowledge in a design challenge, begin investigating a related concept, or delve deeper into the current concept. The goal of the expand phase is to deepen knowledge. In our inquiry units students deepen and extend their knowledge in various ways, such as investigating the adaptations of local birds, exploring the work of paleontologists, selecting trees for a park, and designing a container that prevents ice from melting.

- **Assess:** The assess phase is embedded throughout the learning cycle. This allows for both formative and summative assessment. Formative assessments are typical in the engage and explore phases in order to monitor student progress and make needed instructional modifications. The explain phase always has a summative assessment in which students demonstrate understanding, but can also include formative assessment. The expand phase may include either formative or summative assessment, depending on the nature of the expand activities. Some of the formative assessments used in our inquiry units are guiding questions, think-pair-share, Venn diagrams, science notebook entries, and formative assessment probes.

STRENGTHS AND LIMITATIONS OF THE LEARNING CYCLE

The learning cycle is a widely accepted model for planning inquiry instruction. A robust body of research indicates that science instruction using learning cycle models results in the development of more sophisticated scientific reasoning, an increased interest in science content, and an increased achievement in the learning of the content (Bybee et al. 2006).

Brown and Abell (2007) found that hands-on activities are a necessary but insufficient means of teaching science. If teachers do not work through an entire learning cycle with their students, student learning suffers. Roth (as cited in Brown and Abell 2007) found that fifth-grade students understood scientific concepts better when exploration was followed with discussion and writing. The discussion and writing occur in the explain phase, after the hands-on experiences common to the explore phase. There is sometimes a tendency to rush through learning cycle lessons or units. Skipping phases or giving in to the temptation to explain the concepts to students during the explore phase results in less satisfactory results. Good science instruction—particularly when it includes reading, writing, illustrating with drawings, graphs, and charts, and the development of products—takes time.

Another limitation arises when instructional planning focuses on identifying "E" activities without careful consideration of the learning that should be taking place in each phase. The Es in the 5E learning cycle are easy to remember. But we've seen many "learning cycle" lessons that were a disjointed series of hands-on activities. For the learning cycle to be effective, the activities must be a coherent, carefully structured set of experiences that allow students to construct meaning by linking evidence and explanations.

There is an undeniable link between inquiry and the learning cycle. The learning cycle provides a simple yet powerful framework for implementing inquiry instruction that spans a lesson, a unit, or a year.

REFERENCES

Brown, P. L., and S. K. Abell. 2007. Examining the learning cycle. *Science and Children* 44 (5): 58–59.

Bybee, R. W., J. A. Taylor, A. Gardner, P. Van Scotter, J. C. Powell, A. Westbrook, and N. Landes. 2006. *The BSCS 5E instructional model: Origins, effectiveness, and applications.* Colorado Springs, CO: BSCS.

Center for Science, Mathematics, and Engineering Education. 2000. *Inquiry and the national science education standards: A guide for teaching and learning.* Washington, DC: National Academies Press.

Chessin, D. A., and V. J. Moore. 2004. The 6-E learning model. *Science and Children* 42 (3): 47–49.

Duschl, R. A., A. H. Schweingruber, and A. W. Shouse. 2007. *Taking science to school: Learning and teaching science in grades K–8.* Washington, DC: National Academies Press.

Eisenkraft, A. 2003. Expanding the 5E model. *Science Teacher* 70 (6): 56–59.

Hapgood, S., S. J. Magnusson, and A. S. Palincsar. 2004. Teacher, text, and experience: A case of young children's scientific inquiry. *Journal of the Learning Sciences* 13 (4): 455–505.

Karplus, R., and R. G. Fuller. 2002. *A love of discovery: Science education, the second career of Robert Karplus.* New York: Kluwer Academic/Plenum Publishers.

Martin, R., C. Sexton, K. Wagner, and J. Gerlovich. 1997. *Teaching science for all children.* Needham Heights, MA: Allyn and Bacon.

Ohio Resource Center. n.d. *The learning cycle. http://ohiorc.org/pm/science/Sci_LearningCycle. aspx*

Chapter 3
Authentic and Relevant Literacy Experiences in Inquiry

The learning cycle provides many opportunities to develop and refine literacy skills. Best practices in literacy (Gambrell, Malloy, and Mazzoni 2011) include creating a classroom culture that fosters literacy motivation. One way to accomplish this is to make all aspects of literacy (reading, writing, speaking, listening, and viewing) authentic and relevant to students. Authentic literacy activities mirror those that happen in people's everyday lives, as opposed to only in the context of the classroom (Purcell-Gates 2002). For example, reading a cookbook to learn how to prepare a new recipe is an authentic event; reading to answer questions on a worksheet is not. In one study, teachers who incorporated authentic literacy events into their instruction had students that demonstrated greater growth in comprehension and writing (Duke et al. 2006). In "Let's Dig!" (Chapter 14), students read to learn how paleontologists excavate fossils; they then write their own work plans to conduct simulated fossil digs in the classroom. In "Measuring Pennies and More" (Chapter 7), students read and discuss units and tools of measurement and then apply this information as they design and write an experimental procedure. In both cases, the literacy activities reflect those performed in real life: reading to learn how a task is accomplished, discussing that information with others, and writing a procedure to be carried out.

In creating our units, we strove to make the science concepts and associated literacy activities relevant to students, beginning with the engage phase. For example, in "Classroom Curling" (Chapter 9), students play games of curling and are invited to think about strategies for scoring points. In "My Favorite Tree" (Chapter 11), students take a nature walk to observe trees. Later, they examine the characteristics of various tree species, ideally those observed on the walk or around their school and homes. By situating the ensuing science investigations and literacy tasks within the context of shared student experiences, we create relevance and increase engagement and motivation.

Each unit is built around a multigenre nonfiction text set—a collection of texts that present a robust view of science and of particular science concepts. Throughout the learning cycle, students interact with the texts in a variety of ways: listening to read-alouds in whole-class and small-group settings and independent reading. As students listen and read, they are able to connect concepts between texts, as well as to their prior knowledge or schema, to construct meaning (Hartman 1995). In the explore phase of "Beaks and Biomes" (Chapter 10), students each select, read, and discuss a text of their own choosing as part of an idea circle to learn about the tundra environment. In "Scientists Like Me" (Chapter 6), students listen to and read a variety of biographies to identify scientific behaviors and habits of mind. The construction of knowledge across multiple texts is an essential component of the units. Using multiple texts that build on prior knowledge, link concepts, and expand vocabulary is another best practice supported by literacy research (Gambrell, Malloy, and Mazzoni 2011).

Within each unit, literacy activities seamlessly blend with the science content and inquiry experiences. Reading, writing, speaking, listening, and viewing not only support the development of science concepts, but are important literacy experiences in their own right.

READING

Throughout the units, students read for specific purposes. They read to learn more about what they've observed in hands-on explorations. In the explore phase of "Minds-on Matter" (Chapter 8), students read about each state of matter after making careful observations. They also compare information from the text to their observations, adding new evidence to what they've already noted.

Text is also used when firsthand observation or investigation is impractical or insufficient to create an understanding of the scientific concepts. In "Patterns in the Sky" (Chapter 15), students make observations of shadows, the phases of the Moon, and constellations using firsthand experiences and technology. However, these observations are not enough to create an understanding of why these patterns occur. Students turn to texts to build on these observations and clarify their learning.

In "Classroom Curling," students construct definitions for key vocabulary terms by drawing upon the data from their investigations and information from several texts. Reading provides a basis for rich discussion in the idea circle activity in "Beaks and Biomes" and the Seed Discussion in "Measuring Pennies and More."

Reading in the context of the learning cycle also provides opportunities for students to practice reading strategically. In recent years, most comprehension instruction has focused on the use of seven research-based cognitive and metacognitive skills, or reading strategies: predicting, using background knowledge to make connections, setting purposes for reading, visualizing, identifying text structure, monitoring comprehension, and summarizing. The strategies are used as a text is read to facilitate understanding (Almasi et al. 2011). These strategies were initially taught one at a time and in isolation, but research suggests that it is more effective to present students with a set of strategies that can be used flexibly (Pressley et al. 1995; Schuder 1993). We do not explicitly teach strategies in our inquiry units, but we certainly encourage you to reference them and invite students to apply them as needed. Strategies might be introduced and discussed during reading instruction, and then applied while reading science-themed text. Doing so not only allows students to practice applying strategies in varied contexts but also builds comprehension of challenging informational texts.

WRITING

Writing is also used throughout each learning cycle unit. Students take notes and use graphic organizers like Venn diagrams to organize their thoughts as they construct meaning from their investigations. They write to define a procedure to be followed, as in "Let's Dig!" They write to define vocabulary, as in "Classroom Curling." They write to communicate understanding and demonstrate knowledge, as in the explain phase of each unit. Whenever possible, students write for audiences besides their teachers. Writing for authentic audiences and purposes produces more effective writing (Slagle 1997).

Drawing and illustrating are also forms of writing. Students draw the fossils they discover in "Let's Dig!" They also engage in a form of writing when they construct *infographics,* which are diagrams, charts, tables, and graphs that combine visual information and text. This is the case in "Drip Drop Detectives" (Chapter 13), when students work collaboratively to produce a diagram illustrating the water cycle; in "My Favorite Tree," as students create a chart illustrating similarities and differences between tree species; and in "Let's Dig!" when students create cross sections of their dig sites. Creating infographics requires a high level of understanding, making it an effective way for students to demonstrate their knowledge. It is particularly effective for students who struggle to express themselves in writing. As Steve Moline (1995, p. 1) notes, "students who are judged to be 'poor writers' are sometimes discovered to be excellent communicators if they are allowed the option to write the same information in a visual form."

SPEAKING

Each inquiry unit provides frequent opportunities for students to engage in conversation as a whole class, in small groups, and in pairs. They discuss the observations they've made, the data they've collected, and the texts that they have read. Some conversations are teacher directed; others are student directed. Balancing these two types of discussions is another research-based best practice in literacy (Gambrell, Malloy, and Mazzoni 2011), but it applies equally to science instruction. These discussions generate communal knowledge, as in the case of the idea circle activity in "Beaks and Biomes." They provide an opportunity for students to organize and clarify their thinking, and an opportunity to share their findings with others, as in "Measuring Pennies and More" when students prepare and perform an interview about a new discovery.

Research demonstrates that discussions around text promote not only comprehension but also higher-order thinking (Almasi, McKeown, and Beck 1996; Almasi, O'Flahavan, and Arya 2001). Peer discussion has also been found to enhance student understanding of science concepts (Smith et al. 2009).

LISTENING

Listening comprehension is an underemphasized, yet important, component of literacy instruction. Opitz and Zbaracki (2004) remind us that listening is a complex process that is different from hearing. While hearing only involves discriminating between sounds, listening implies processing information and comprehension. Read-alouds are an important tool for promoting listening comprehension. When students listen to a read-aloud, they are actively connecting the new information to their prior experience. It also gives students a chance to work with challenging text and concepts without being burdened by decoding demands. All of our inquiry units include read-alouds and opportunities for students to develop listening comprehension through teacher-directed and student-directed discussions.

VIEWING

In today's technology-driven society we are bombarded with infographics and information, making the ability to gather information and make meaning from visual images extremely important. As with reading, writing, speaking, and listening, students need explicit instruction, modeling, and guided practice to be successful with infographics. Our inquiry units provide opportunities for students to practice interpreting visual information. In "Drip Drop Detectives," students examine diagrams depicting the water cycle before creating their own. In "Come Fly With Me" (Chapter 12), students make predictions about how birds fly and then watch video clips of the birds in flight to either confirm or refute their predictions. Just as creating infographics is an effective strategy for struggling writers, interpreting infographics is an effective strategy for struggling readers. Information presented in this way is accessible to all students, including English-language learners (Moline 1995).

We have integrated inquiry and literacy in learning cycle units in authentic and relevant ways. Through these units students have opportunities to develop scientific content knowledge and processes while engaging in authentic literacy activities that expand and hone their reading, writing, speaking, listening, and viewing skills.

REFERENCES

Almasi, J. F., M. G. McKeown, and I. L. Beck. 1996. The nature of engaged reading in classroom discussions of literature. *Journal of Literacy Research* 28: 107–146.

Almasi, J. F., J. F. O'Flahavan, and P. Arya. 2001. A comparative analysis of student and teacher development in more proficient and less proficient peer discussions of literature. *Reading Research Quarterly* 36: 96–120.

Almasi, J. F., B. M. Palmer, A. Madden, and S. Hart. 2011. Interventions to enhance narrative comprehension. In *Handbook of reading disability research*, eds. R. Allington and A. McGill-Franzen, 329–344. New York: Routledge.

Duke, N. K., V. Purcell-Gates, L. A. Hall, and C. Tower. 2006. Authentic literacy activities for developing comprehension and writing. *The Reading Teacher* 60: 344–355.

Gambrell, L. B., J. A. Malloy, and S. A. Mazzoni. 2011. Evidence-based best practices in comprehensive literacy instruction. In *Best practices in literacy instruction*, 4th ed., ed. L. M. Morrow and L. B. Gambrell. New York: Guilford.

Hartman, D. K. 1995. Eight readers reading: The intertextual links of proficient readers reading multiple passages. *Reading Research Quarterly* 30: 520–561.

Moline, S. 1995. *I see what you mean: Children at work with visual information*. Portland, ME: Stenhouse.

Opitz, M. F., and M. D. Zbaracki. 2004. *Listen hear! 25 effective listening comprehension strategies*. Portsmouth, NH: Heinemann.

Pressley, M., R. Brown, P. Van Meter, and T. Schuder. 1995. Transactional strategies. *Connecting With the Community and the World of Work* 52 (8): 81–82. *www.ascd.org/publications/educational-leadership/may95/vol52/num08/-Transactional-Strategies.aspx*

Purcell-Gates, V. 2002. Authentic literacy in class yields increase in literacy practices. *Literacy Update* 11 (7): 9. *www.lacnyc.org/resources/publications/update/Update2001-02/Update02-05.pdf*

Schuder, T. 1993. The genesis of transactional reading instruction in a reading program for at-risk students. *The Elementary School Journal* 94: 183–200.

Slagle, P. 1997. Getting real: Authenticity in writing prompts. *The Quarterly* 19 (3): 20–23. *www.nwp.org/cs/public/print/resource/882*

Smith, M. K., W. B. Wood, W. K. Adams, C. Wieman, J. K. Knight, N. Guild, and T. T. Su. 2009. Why peer discussion improves student performance on in-class concept questions. *Science* 323 (5910): 122–124. *www.sciencemag.org/content/323/5910/122.short*

Chapter 4
Nonfiction Text Sets

I was an avid reader as a child. My mother nurtured my love of reading through weekly trips to the library and a subscription to the magazine *My Big Backyard*. She read the articles to me until I was old enough to read them myself. I eventually graduated to *Ranger Rick*, and loved reading about interesting animals and the places they lived. I distinctly remember reading about the recovery of the area around Mount Saint Helens in the years following the 1980 eruption and marveling at the huge trees scattered like toothpicks in a photograph of the area. Many years later I would visit Mount Saint Helens, and while driving and hiking through the park, the photograph from that article was always in my mind's eye.

Recently, I observed a fifth-grade reading class. The teacher opened the lesson by introducing a nonfiction picture book. She asked the students to share what types of nonfiction they liked to read. One girl blurted out, "Books about Justin Bieber!" Two others shared that they liked to read about dogs. Another raised her hand and volunteered, "This may sound funny, but sometimes I like to flip through the encyclopedia and read about different things."

Children are inherently curious about the world around them, and they enjoy reading about it. Whether it's about animals, trucks, dinosaurs, or planets, nonfiction text is a natural fit. As teachers we have the opportunity to harness that curiosity and help students find books that they enjoy reading.

—Jessica

Literacy instruction has changed in recent years to place a greater emphasis on nonfiction text with students of all ages. Researchers and teachers have realized that reading nonfiction can build important background knowledge (Soalt 2005) and vocabulary (Duke, Bennett-Armistead, and Roberts 2003). These benefits seem particularly pronounced in content areas such as science. One study of an integrated instructional approach called concept-oriented reading instruction (CORI) found that when combined with hands-on investigations, reading and listening to informational text was a valuable tool for knowledge building (Guthrie et al. 1999).

Nonfiction text is also more challenging for students to read, because of the text structures (McCormick 1995), specialized vocabulary, and high density of concepts (Merkley and Jefferies 2000). Captions, keys, and other text features also need to be considered while reading. Students need explicit instruction in key vocabulary terms and in text structures such as cause and effect, compare and contrast, problem and solution, and sequence classification (McCormick 1995). Within a learning cycle approach, this explicit instruction is best suited for the explain phase, after students have had time to develop firsthand experience during the explore phase. Without sufficient instructional time devoted to this genre, students will not be successful with this important type of text.

A great deal of reading and writing in everyday life is nonfiction. Children encounter nonfiction text when they follow the directions to assemble a model, find out how to get to the next level of a video game, or bake cookies. Proficiency with nonfiction text is also key to success in later schooling. For these reasons, there has been a growing recognition that children's classroom experiences should include greater exposure to this type of text. Some students prefer nonfiction text, so increasing classroom use of the genre may help motivate them to read (Caswell and Duke 1998). The beautifully illustrated and well-written nonfiction children's books available today increase the appeal of this genre to all students.

NONFICTION GENRES

Nonfiction text sets are integral to our approach to integrating science and literacy instruction. A well-constructed text set includes multiple nonfiction genres. Ebbers (2002) describes seven genres of nonfiction text: reference, explanation, narrative expository, how-to, biography, field guide, and journal. To this set we add poetry, as many excellent examples of science-themed poetry are available. For each genre we provide a description and examples of published children's literature (see the "Children's Literature" reference list at the end of this chapter for publication details).

Reference

Reference texts provide accounts of phenomena based on current scientific understanding (Ebbers 2002). Encyclopedias are a type of reference text, but others exist as well. One example is *Into the Air: An Illustrated Timeline of Flight* by Ryan Ann Hunter (2003). This book provides an illustrated history of flight, from the first flying creatures on Earth to modern-day air and space travel. Another is *What Do You Know About Fossils?* by Suzanne Slade (2008). The book asks and answers 20 questions about fossils.

Explanation

Explanation texts do just what their name suggests: explain how something happens or why it occurs (Veel 1993, as cited in Ebbers 2002). Consider *Exploring Forces and Movement* by Carol Ballard (2008), which includes this explanation in the "Friction" section:

> Whenever two surfaces rub against each other, there is a force called friction.
> Even surfaces that are very smooth create friction. As one surface moves over
> the other, the rough bits catch on each other and cause friction. The rougher
> the surfaces, the more friction there will be. (p. 18)

Other explanation texts include *Birds: Nature's Magnificent Flying Machines* by Carol Arnold (2003) and *Beaks!* by Sneed B. Collard III (2002). These texts explain how birds fly and how their beaks are adapted for specific food sources.

Narrative Expository

Narrative expository texts relate factual information through a familiar structure of a story (Ebbers 2002). For example, *The Snowflake: A Water Cycle Story* by Neil Waldman (2003) traces the path of water through various forms and across the world:

MARCH

As the sun grew warmer, the ice began to melt. The snowflake became a droplet of water once again. It fell through a crack in the rocky pond bottom and trickled down into the ground. Downward it sank, into the blackness within the mountain. Along with millions of other droplets, it splashed into an underground stream that flowed deep into the earth. (unpaged)

Other examples of narrative expository texts include *Newton and Me* by Lynne Mayer (2010), which tells the story of a young boy exploring forces and motion. *Measuring Penny* by Loreen Leedy (1997) recounts the story of a young girl who creatively uses measurement as part of a class assignment.

How-To

How-to texts also do what their name implies: describe how to conduct an investigation, complete a task, or construct an object. How-to guides provide procedural steps in words and/or illustrations or photographs. We include *The Kids' Guide to Paper Airplanes* (Harbo 2009) in the "Come Fly with Me" inquiry unit (Chapter 12) to help students construct two types of paper airplanes.

Biography

Biographies detail the life story of an individual. They may recount the person's entire life or just a portion of it. Today's biographies for elementary students are often in picture book format, with beautiful illustrations and rich, engaging text. Biographies mentioned in this book include *The Boy Who Drew Birds: A Story of John James Audubon* by Jacqueline Davies (2004) and *Into the Deep: The Life of Naturalist and Explorer William Beebe* by David Sheldon (2009).

Field Guide

Field guides, as their name suggests, are books that help scientists and other individuals identify and classify objects (Ebbers 2002) when they encounter them in nature. They provide information about the characteristics of natural objects, such as animals, plants, rocks, minerals, and fossils. Detailed illustrations or photographs make features visible to students, and the text introduces them to the scientific terminology used for objects they see all around them. In the "Come Fly with Me" unit (Chapter 12), students use field guides like *Birds of North America* (Robbins, Bruun, and Zim 2001).

Journal

Journals relate procedural information or day-to-day happenings in a narrative format (Ebbers 2002). A journal might recount a scientist's experience in the field or a procedure completed in a laboratory. For example, in *My Season With Penguins: An Antarctic Journal* (2000), Sophie Webb notes:

> 20 December
>
> Today we attach radio transmitters to fifteen penguins. Each transmitter is tuned to a unique frequency. This will allow us to track individual penguins when they go out to sea to feed….The adult penguins with chicks can be extremely defensive, which makes them difficult to catch without getting painful bruises. (p. 23)

Other examples of journals include Webb's second book, *Looking for Seabirds: Journal From an Alaskan Voyage* (2004). *Close to the Wind: The Beaufort Scale* by Peter Malone (2007) also incorporates a journal throughout the narrative expository text.

Poetry

Poetry presents scientific content in a unique and engaging format. Whether rhymed text, free verse, or shape poetry, the genre can capture students' interest and appeal to their creativity. *Faces of the Moon* by Bob Crelin (2009) uses rhymed text to describe the phases of the Moon:

> For as she orbits Earth in space,
> the sunlight moves across her face,
> and from our world, beneath the sky,
> we watch Moon change as days go by.
>
> From night to night and day to day,
> she'll wax (or grow), then wane away.
> Each phase has earned a name we say
> as light and shadows change. (unpaged)

Another example of science-themed poetry is *S Is for Scientists: A Discovery Alphabet* by Larry Verstraete (2010), used in "Scientists Like Me" (Chapter 6).

Although these genres provide a useful way to think about nonfiction texts, the lines between them are often blurred. Additionally, some books contain multiple genres. For example, books in the *Investigate Science* series by Melissa Stewart (which includes titles such as *Air is Everywhere* [2004], and *Down to Earth* [2004]) blend reference, explanation, and how-to-text. While these books complicate classification, they serve a multitude of purposes in the science classroom and are worth including.

RELEVANCE OF GENRES TO SCIENTIFIC INQUIRY AND THE LEARNING CYCLE

When we consider science-themed nonfiction text, each genre sheds light on a particular aspect of scientific inquiry (Ebbers 2002). Reference and explanation books share current scientific understandings, and when older titles are considered, demonstrate that these understandings can change in light of newly discovered evidence. Narrative expository texts help us understand phenomena by relating the events to familiar story structure. How-to texts speak to the need for scientific investigations to be replicable. Biographies remind us that science is a human endeavor, and that men and women throughout history and across the world have participated (and continue to participate) in the activities known as science. Field guides explore the world's diversity through observable characteristics—a key feature of scientific inquiry. Journals share the true nature of science—what it is actually like to be in the field or in the laboratory. And while poetry wasn't included in Ebbers's list of genres, we feel that when it communicates accurate science, its creative format and engaging text makes it worthy of inclusion in nonfiction text sets.

The different nonfiction genres also lend themselves to different phases of the learning cycle. Everett and Moyer (2009) suggest the following guidelines for selecting books for each phase:

- *Engage: Use books that help generate questions.* Narrative expository and poetry are often perfectly suited to pique student interest.

- *Explore: Use books that pose readily testable questions; books that are read and then set aside during the investigation.* How-to books help students plan the investigations they will carry out, and field guides help them identify and classify living organisms, rocks, and fossils. Reference and explanation books also are essential for the explore phase, as they provide a source of information with which students can compare their own findings.

- *Explain: Use books that address the learning objectives, sometimes reading only the pertinent parts.* In our approach, we use these types of books (typically reference and explanation) at the end of the explore phase. In our explain phases, we suggest titles for use as mentor texts (Dorfman and Cappelli 2009), examples of format and writing style after which students can pattern their own work. A variety of nonfiction genres serve this purpose.

- *Extend: Use books that help students make real-world connections or connect to other concepts.* This is similar to our expand phase, in which any number of genres might be used. Journals and biographies expand students' understanding of a topic by introducing them to men and women involved in a particular field. Reference and explanation provide additional information to extend student learning. Narrative expository texts and poetry provide a creative twist on the information and bridge to related concepts.

- *Evaluate: Use books to pose questions or present problems on topics that students can respond to in writing, through illustrations, or through integrated text (a combination of visual and written information).* This is similar to our assess phase. We do not use texts during the assess phase, because the summative assessment tasks have been completed during the explain phase.

Our suggestions for matching learning cycle phases and nonfiction text genres are illustrated in Figure 4.1.

FIGURE 4.1. LEARNING CYCLE PHASES AND NONFICTION TEXT GENRES

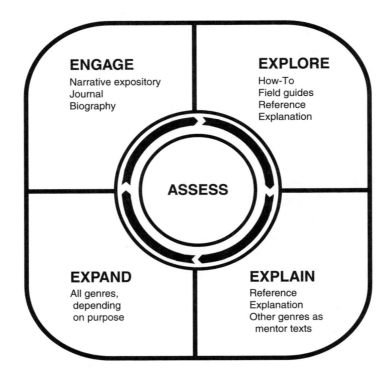

The guidelines described by Everett and Moyer (2009) and in Figure 4.1 are not hard-and-fast rules, and a creative teacher will find ways to use a given text in just about any context. However, in developing the units for this book, we found ourselves drawn to certain genres over and over again for particular contexts within the units.

While each genre of nonfiction and/or informational text conveys an important type of knowledge, the real benefit can be obtained from pairing or grouping them. When students are only exposed to reference and explanation books, they come to view science as a disjointed collection of facts. But when these genres are read in tandem, they act synergistically to create a robust understanding of the nature of science. As a result, students are exposed to a more accurate view of science and how scientific knowledge is obtained.

TEXT SETS

As defined by Short, Harste, and Burke (1996), a text set is a collection of 5–15 books that relate conceptually in some way. Although text sets may be fiction or nonfiction and may be organized around any content area, we focus on nonfiction sets that relate to science content. We've also slightly modified the definition of a text set by dropping the requisite number of books, because the availability of high-quality texts varies greatly across the scientific disciplines. The number of books in each text set in the inquiry units (Chapters 6–15) range from 4 to 12.

In Part II, we present inquiry units that involve multigenre nonfiction text sets—sets that include a variety of the genres described in this chapter. For example, the text set used in "Let's Dig!" (Chapter 14) includes *Mary Anning and the Sea Dragon* (biography), *Dinosaur Dig!* and *Mysteries of the Fossil Dig: How Paleontologists Learn About Dinosaurs* (reference), *Dinosaur Mountain: Digging Into the Jurassic Age* (narrative expository), and *Fossils* (field guide). These books (and others) combine to support an inquiry-based learning cycle in which students develop and conduct a fossil dig to answer the question, *What can we learn from studying fossils?*

A common approach to bringing nonfiction text into the classroom is to simply check out every book on a topic from the library. While clearing the shelves can be useful, particularly in research projects, this is not the same as creating a text set. The latter involves a thoughtful and purposeful selection of a small number of texts based on their content, genre, and intended use in teaching and learning. Each book in a text set is included for a specific reason. Focusing on a limited number of texts allows you to be much more thoughtful and purposeful in how you use them in your instruction.

We should note that a text set need not include all eight genres described here. Again, the diversity of texts will vary based on your objective and the subject matter. If you've spent much time reviewing science-themed children's literature, you know that certain subjects, like animals, dinosaurs, and plants, are much better represented than physical science topics like forces and motion or gravity. However, we feel that a true text set should represent the diversity of nonfiction texts, so at least three genres is a good guideline.

The rewards of creating nonfiction text sets are great for both teachers and students. Teachers we have worked with say that this approach has increased their understanding of nonfiction text and enhanced their awareness of the great diversity within the broad category of nonfiction text. They also say that it has diversified the types of texts they use in their instruction. And, as you'll see in Part II of this book, nonfiction text sets are perfectly suited to support inquiry-based instruction.

CHILDREN'S LITERATURE

Arnold, C. 2003. *Birds: Nature's magnificent flying machines*. Watertown, MA: Charlesbridge.

Ballard, C. 2008. *Exploring forces and movement*. New York: Rosen.

Collard, S. III. 2002. *Beaks!* Watertown, MA: Charlesbridge.

Crelin, B. 2009. *Faces of the moon*. Watertown, MA: Charlesbridge.

Davies, J. 2004. *The boy who drew birds: A story of John James Audubon*. New York: Houghton Mifflin.

Harbo, C. L. 2009. *The kids' guide to paper airplanes*. Mankato, MN: Capstone Press.

Hunter, R. A. 2003. *Into the air: An illustrated timeline of flight*. Washington, DC: National Geographic Society.

Leedy, L. 1997. *Measuring Penny*. New York: Henry Holt.

Malone, P. 2007. *Close to the wind: The Beaufort scale*. New York: The Penguin Group.

Mayer, L. 2010. *Newton and me*. Mt. Pleasant, SC: Sylvan Dell.

Robbins, C. S., B. Bruun, and H. S. Zim. 2001. *Birds of North America*. New York: St. Martin's Press.

Sheldon, D. 2009. *Into the deep: The life of naturalist and explorer William Beebe*. Watertown, MA: Charlesbridge.

Slade, S. 2008. *What do you know about fossils?* New York: Rosen.

Stewart, M. 2004. *Air is everywhere*. Minneapolis, MN: Compass Point Books.

Stewart, M. 2004. *Down to Earth*. Minneapolis, MN: Compass Point Books.

Verstraete, L. 2010. S *is for scientists: A discovery alphabet*. Ann Arbor, MI: Sleeping Bear Press.

Waldman, N. 2003. *The snowflake: A water cycle story*: Brookfield, CT: Millbrook Press.

Webb, S. 2000. *My season with penguins: An Antarctic journal*. Boston, MA: Houghton Mifflin.

Webb, S. 2004. *Looking for seabirds: Journal from an Alaskan voyage*. Boston, MA: Houghton Mifflin.

REFERENCES

Caswell, L. J., and N. K. Duke. 1998. Non-narrative as a catalyst for literacy development. *Language Arts* 75 (2): 108–117.

Dorfman, L. R., and R. Cappelli. 2009. *Nonfiction mentor texts: Teaching informational writing through children's literature, K–8*. Portland, ME: Stenhouse.

Duke, N. K., V. S. Bennett-Armistead, and E. M. Roberts. 2003. Bridging the gap between learning to read and reading to learn. In *Literacy and young children: Research-based practices*, eds. D. M. Barone and L. M. Morrow, 226–242. New York: Guilford Press.

Ebbers, M. 2002. Science text sets: Using various genres to promote literacy and inquiry. *Language Arts* 80 (1): 40–49.

Everett, S., and R. Moyer. 2009. Literacy in the learning cycle: Incorporating trade books helps plan inquiry-learning experiences. *Science and Children* 47 (2): 48–52.

Guthrie, J. T., E. Anderson, S. Alao, and J. Rinehart. 1999. Influences of concept-oriented reading instruction on strategy use and conceptual learning from text. *The Elementary School Journal* 99 (4): 343–366.

McCormick, S. 1995. *Instructing students who have literacy problems*. Englewood Cliffs, NJ: Prentice Hall.

Merkley, D. M., and D. Jefferies. 2000. Guidelines for implementing a graphic organizer. *The Reading Teacher* 54 (4): 350–357.

Short, K. G., J. C. Harste, and C. L. Burke. 1996. *Creating classrooms for authors and inquirers*. Portsmouth, NH: Heinemann.

Soalt, J. 2005. Bringing together fictional and informational texts to improve comprehension. *The Reading Teacher* 58 (7): 680–683.

Chapter 5
Getting Started With the Inquiry Units

t is always exciting to try something new in your classroom, but we know it can also be challenging. Here are some things to think about as you prepare to teach the integrated inquiry units. Step one is, of course, to read the unit beginning to end. Here are some more.

GATHER THE BOOKS AND MATERIALS

Go to the library several weeks before you begin the unit. Your local library may not have all of the titles we have suggested, but you may be able to borrow the books through an interlibrary loan program. You may also be able to find good used copies at a secondhand bookstore. There is always the option of purchasing the books if necessary. If you substitute other titles, select books that are well suited for the intended purpose. Carefully verify the scientific accuracy of the content and the reading level. We've included the Flesch-Kincaid reading level, which expresses the difficulty of a text as a grade level. For example, a Flesch-Kincaid reading level of 4.5 indicates that the text is suitable for reading by students in fourth grade. Other methods of calculating reading levels of text (e.g., the Lexile Framework) exist, which may be useful if you substitute other titles for those presented in the unit.

The materials required for the inquiry units are common items that you may already have in your classroom. If not, they can be found at most discount department stores.

Occasionally online resources are used in the inquiry units. We always provide the URL for such resources. We have also provided QR codes for your convenience. These codes allow you to use a scanning app on your smartphone, on your tablet, or with the webcam on your computer to scan and quickly access the resources. Some of these online resources (including videos from YouTube) may not be accessible on your school's servers. If so, you can work around this by downloading the videos on a computer at home and saving the files to a USB drive. The videos will be accessible without connecting to the YouTube site directly.

CREATE A SUPPORTIVE CLASSROOM ENVIRONMENT

The units require collaboration, communication, and firsthand investigation. We anticipate that your classroom will be a beehive of activity as students eagerly explore and inquire. Plan in advance for how you will manage materials and a high level of activity while you are facilitating learning. Communication is key to the successful completion of the units. Expect teacher-student, student-teacher, and student-student discussions to be an ongoing and critical part of the learning environment. Establish classroom norms for working under such conditions up front so the students know what is expected. Work with your students to establish these norms. They are more likely to remember and adhere to expected behaviors when they have had a role in setting the expectations.

An important part of the classroom environment is the comfort level students feel when sharing their thinking. Craft discussions so that students are confident that you want to know what they are thinking as opposed to feeling the pressure to provide a correct answer.

PREPARE GUIDING QUESTIONS TO ENCOURAGE THINKING

A testable question is introduced early in each unit. The testable question frames the inquiry. Guiding questions support students as they work to understand the relevant concepts. Most of these questions do not have right or wrong answers. They are designed to foster discussion and the open sharing of ideas. Questions of this nature allow you to peek in on your students' thinking. Their responses will expose more about what students know than any multiple-choice end-of-the-chapter questions ever could. Listen closely to what your students are saying. They are just as likely to show depth of understanding as they are to uncover misconceptions. In each unit we have provided sample guiding questions. As you get ready to teach the units, prepare additional questions for follow-ups to the ones we have provided or rephrase ours if necessary. Questions are critical to inquiry in more than one way!

INTEGRATE ACROSS DISCIPLINES

Our units are written to promote conceptual development. Research tells us that students need to experience the same information three to four times before incorporating it into their worldview (Nuthall 1999). Olson (2009) tells us that our instruction should include repeated cycles of exploration and concept development to move students toward the "big idea." These principles informed our writing of the units, and as a result the units may seem long. However, because of our integrated approach, the activities may be incorporated in both science and English language arts class periods. As you read the unit, consider which activities are best suited for science class and which can be completed in English language arts. In general, introducing the units and testable questions should be done in science. Activities that involve hands-on explorations or designing and conducting experiments also belong in science. Those that involve reading, discussion around text, or writing can be completed in English language arts.

USE FORMATIVE AND SUMMATIVE ASSESSMENT

As you and your students work through the inquiry units, you will find that we focus much more on formative than summative assessment. Like many others we believe that formative assessment informs instruction and leads to better outcomes. Only formative assessments are used in the engage and explore phases, while students are developing understanding. Formative assessments allow you to collect evidence of student learning and to use that evidence to make instructional decisions. You may find that students need extra support to complete some tasks or that they can work independently earlier than you expected. Either way it is important to adjust your instruction accordingly. Formative assessments are typically not graded. The purpose of formative assessment is to gather data to inform instructional decisions.

Summative assessments are typically found in the explain phase of the inquiry units. The purpose of summative assessment is to evaluate student learning. This is typically where a grade is assigned. We have provided two rubrics for your use. One is for assessing content knowledge and literacy skills, the other is for assessing science process skills. The Science and Literacy Rubric found in Appendix 2 (p. 276) is meant to be used with the summative assessments found in the explain phase and occasionally in the expand phase. In our experience rubrics are extremely helpful for identifying and describing the desired outcomes. However they do not always translate well into letter grades. We have adapted Fanning and Schmidt's (2007) achievement grading approach for use with the units. This approach combines a rubric with a method for assigning a letter grade. The rubric and the achievement grading scale can be found in Appendix 2. Appendix 2 also includes the Science Process Skills Rubric, a formative assessment rubric that can be used intermittently throughout the academic year; a student version of the rubric is also found in Appendix 2.

UNITS AT A GLANCE

Table 5.1 (pp. 38–42) summarizes the objectives, number of class periods, text set titles, and materials needed for each inquiry unit.

REFERENCES

Fanning, M., and B. Schmidt. 2007. Viva la revolución: Transforming teaching and assessing student writing through collaborative inquiry. *English Journal* 97 (2): 29–35.

Nuthall, G. 1999. The way students learn: Acquiring knowledge from an integrated science and social studies unit. *Elementary School Journal* 99 (4): 303–341.

Olson, J. K. 2009. Methods and strategies: Being deliberate about concept development. *Science and Children* 46 (6): 51–55.

TABLE 5.1. INQUIRY UNITS SUMMARY

Chapter Number and Title	Objectives	Class Periods (45 min.)	Text Set Titles	Materials
Chapter 6 Scientists Like Me	• Express ideas about scientists in drawings • Name scientists and relate their accomplishments • Identify similarities and differences among scientists and between scientists and themselves • Identify scientific behaviors and habits of mind • Practice the following science process skills: observing, communicating, classifying, and experimenting • Write poetry	15	• *Barnum Brown: Dinosaur Hunter* • *The Boy Who Drew Birds: A Story of John James Audubon* • *Come See the Earth Turn: The Story of Léon Foucault* • *Dear Benjamin Banneker* • *Gregor Mendel: The Friar Who Grew Peas* • *Into the Deep: The Life of Naturalist and Explorer William Beebe* • *Rachel: The Story of Rachel Carson* • *Reader of the Rocks* • *S Is for Scientists: A Discovery Alphabet*	Chart paper, markers, string, yardstick, magnifying glasses, student science journals, small objects for classification, objects to be used as pendulum bobs, masking tape, metersticks, stopwatches, large sheets of white construction paper or poster board, copies of supporting documents
Chapter 7 Measuring Pennies and More	• Explain the importance of measurement in science • Practice using a variety of measuring tools • Conduct an investigation in which measurement is a critical component • Devise an ordinal measurement scale • Write an interview for a newscast and conduct the interview for an audience	9	• *Close to the Wind: The Beaufort Scale* • *How Do You Measure Length and Distance?* • *How Do You Measure Liquids?* • *How Do You Measure Time?* • *How Do You Measure Weight?* • *How Tall, How Short, How Far Away* • *Measuring Penny* • *Tornadoes*	A variety of rulers, scales, and timing devices; graduated cylinders, teaspoons, and measuring cups; string; a variety of objects to be used as pendulum bobs; masking tape; scissors; pipette or eyedroppers; pennies; text set; a variety of items that can be weighed and measured; riddle cards; hint cards; event cards; document camera (optional); copies of supporting documents

Chapter Number and Title	Objectives	Class Periods (45 min.)	Text Set Titles	Materials
Chapter 8 Minds-on Matter: Phase Changes and Physical Properties	• Name and describe physical properties of solids, liquids, and gases • Explain how temperature changes during melting, freezing, and boiling • Explain that energy (heat) changes the physical properties of a substance • Write sentences demonstrating cause-and-effect relationships	9	• *Ice Cream: The Full Scoop* • *Ice to Steam: Changing States of Matter* • *A Look at Glaciers* • *Many Kinds of Matter: A Look at Solids, Liquids, and Gases*	Resealable plastic sandwich bags; whole or 2% milk (or juice); vanilla extract; sugar; paper or plastic cups; plastic spoons; a variety of solids and liquids; helium balloons and regular balloons; word wall or chart paper; markers; crayons and pens; glass containers, such as beakers; containers of a variety of sizes and shapes, including bowls; balance or scale; jelly jar or mason jar; bubble wrap; string or rubber band; student science journals; hot plate; wax paper; eyedropper; books about the polar regions; shoe boxes of identical sizes; cardboard, Styrofoam, cotton balls, newspaper, and other items that could be used as insulators; packing tape; large ice cubes or blocks of ice; temperature probe (optional); copies of supporting documents
Chapter 9 Classroom Curling: Exploring Forces and Motion	• Recognize that forces are required to start, stop, or change the direction of an object's motion • Investigate and define friction • Investigate and define gravity • Use text features to locate information in nonfiction texts • Write and illustrate an explanatory report • Collect and analyze data to make evidence-based claims • Make text-to-self connections with nonfiction text	7–9	• *Exploring Forces and Movement* • *Forces: The Ups and Downs* • *Ice Hockey and Curling* • *Motion* • *Newton and Me* • *What Is a Force?*	Masking tape; chalk, wax pencil, or other nonpermanent writing instrument; furniture sliders; objects to use as bumpers; ramp materials and supports for the ramp; printer paper; wax paper; sandpaper (120 grit); stopwatch; string; snack-size plastic bag; weights; graph paper; copies of supporting documents

Chapter Number and Title	Objectives	Class Periods (45 min.)	Text Set Titles	Materials
Chapter 10 Beaks and Biomes: Understanding Adaptation in Migrating Organisms	• Explain the concept of adaptation as it relates to the sanderling • Describe the shore and tundra environments • Practice the following science process skills: predicting, observing, and collecting and analyzing evidence • Read and discuss informational text to locate information • Write nonfiction text in journal format	6	• *Beaks!* • *The Dance of Life* • *Looking for Seabirds: Journal From an Alaskan Voyage* • *Red Knot: A Shorebird's Incredible Journey*	Tweezers, two 8" square aluminum pans, walnut halves, sand, Spanish moss, a variety of stones and pebbles, a variety of shells, items to represent insects and larvae (e.g., pom-poms, wooden buttons, foam insects), larger items to represent lemmings and other small rodents (e.g., wooden spools), chart paper, markers, a set of books about the tundra, bird field guides, computer with internet access and speakers, interactive whiteboard (optional), copies of supporting documents
Chapter 11 My Favorite Tree	• Recognize that trees are a diverse group of organisms • Compare different species within a genus (e.g., sugar maple and silver maple) • Write and illustrate informative text to convey information about trees • Recognize that the characteristics of some trees make them well suited for specific uses • Use multiple sources of information to verify facts	9	• *My Favorite Tree: Terrific Trees of North America* • *National Audubon Society First Field Guide: Trees* • *Peterson First Guides: Trees* • *Trees, Leaves, and Bark*	Clipboards or science notebooks, paper and crayons for rubbings, Computer with internet connection, digital camera (optional), copies of supporting document
Chapter 12 Come Fly With Me	• Explain the relationship between structure and function in birds' wings • Practice the following science process skills: predicting, observing, and collecting and analyzing evidence • Design and interpret infographics • Write expository text to demonstrate understanding	10	• *Birds: Nature's Magnificent Flying Machines* • Field guides such as *Birds of North America* • *Into the Air: An Illustrated Timeline of Flight* • *The Kids' Guide to Paper Airplanes* • *Leonardo and the Flying Boy*	Paper, cut into 6" squares and 7" × 10½" rectangles; computer with internet access; projector or interactive whiteboard; chart paper and marker; masking tape; metersticks; classroom library with flight-related books (optional); document camera (optional); video camera (optional); copies of supporting documents

Chapter Number and Title	Objectives	Class Periods (45 min.)	Text Set Titles	Materials
Chapter 13 Drip Drop Detectives: Exposing the Water Cycle	• Investigate water cycle processes • Construct a water cycle diagram from information found in nonfiction texts • Write an article explaining the water cycle • Recognize that the Earth's supply of freshwater is limited • Gather and analyze water usage data	7	• *Earth's Water Cycle* • *One Well: The Story of Water on Earth* • *The Life and Times of a Drop of Water* • *The Snowflake: A Water Cycle Story* • *The Water Cycle* by Bobbie Kalman and Rebecca Sjonger • *The Water Cycle* by Frances Purslow • *The Water Cycle* by Marcia Zappa • *The Water Cycle* by Rebecca Olien	Graduated cylinders or measuring cups; water; ice; two plants; desk lamp with incandescent bulb; six small plastic containers with lids; three small shallow plastic containers without lids; Styrofoam plates; three 2 L bottles with the bottoms removed or large transparent plastic bags; four 1 L bottles with the tops removed so they can be inverted and used as funnels and small holes drilled in the lids; plastic wrap; markers; approximately 1 cup each of soil, sand, and gravel; chalk; computer with internet access, digital camera (optional); copies of supporting documents
Chapter 14 Let's Dig! Exploring Fossils	• Make inferences about past environments using fossil evidence • Ask and answer questions about fossils using firsthand evidence and nonfiction text • Write procedural and informational text • Write in letter and e-mail formats • Create drawings and infographics such as maps and diagrams • Participate productively as a member of a collaborative group • Collaborate to write a report • Participate in an oral presentation	10	• *Barnum Brown: Dinosaur Hunter* • *Dinosaur Dig!* • *Dinosaur Mountain: Digging Into the Jurassic Age* • *The Dinosaurs of Waterhouse Hawkins* • Field guides such as *National Audubon Society Field Guide to Fossils: North America* • *Fossils* by Melissa Stewart • *Fossils* by Sally M. Walker • *The Fossil Feud: Marsh and Cope's Bone Wars* • *Fossils Tell of Long Ago* • *Mary Anning and the Sea Dragon* • *Mysteries of the Fossil Dig: How Paleontologists Learn About Dinosaurs* • *What Do You Know About Fossils?*	Aquariums or plastic boxes with lids; sand; soil; aquarium gravel in a natural color; a variety of fossils (may be from an existing collection, purchased, or made); string; tape; small shovels; watercolor paintbrushes; magnifying glasses; trays; buckets, boxes, or aluminum pans; colored pencils; graph paper; document camera (optional); digital camera (optional); copies of supporting documents

Chapter Number and Title	Objectives	Class Periods (45 min.)	Text Set Titles	Materials
Chapter 15 Patterns in the Sky	• Observe and explain the changes in shadows throughout the day • Explain the day/night cycle • Observe the phases of the Moon • Observe the motion of constellations • Generalize from their observations that there is a pattern to the appearance and motion of objects in the sky • Use technology to collect data • Collect data over an extended period of time • Write a dialogue	9 and intermittently as needed for the expand phase	• *Faces of the Moon* • *Orion* • *Starry Messenger: Galileo Galilei* • *The Big Dipper*	Chart paper, tape, sunrise and sunset images, globe, 3" x 5" cards and container to hold them, sidewalk chalk, clipboard, flashlights, balls (basketball size), clay, golf tees, drawing supplies (paper, markers, crayons), computer with internet connection, copies of supporting documents

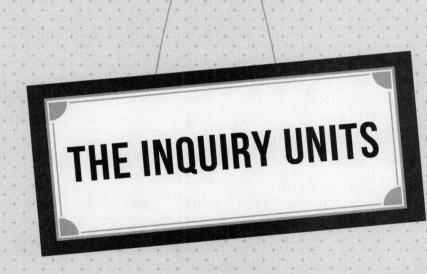

Part II

THE INQUIRY UNITS

Chapter 6
Scientists Like Me

OVERVIEW

In this inquiry, students work to answer two testable questions: *Who can become a scientist?* and *What do scientists do?* Students learn about the lives and work of eight scientists: John James Audubon, Charles William Beebe (known as William Beebe), Rachel Carson, Benjamin Banneker, Barnum Brown, Julie Codispoti, Gregor Mendel, and Jean-Bernard-Léon Foucault (known as Léon Foucault) by reading and discussing picture book biographies. They also practice the science process skills of observing, communicating, classifying, and experimenting (Padilla 1990). Explicit instruction and firsthand experience with process skills is woven into the inquiry at appropriate times so that students experience these behaviors in an authentic way. This pairing breathes life into the behaviors described in the texts and helps students link what they do in the classroom to the work of actual scientists.

This inquiry supports literacy instruction by including several reading strategies. First, students explore the similarities and differences between two scientists, among the entire group of scientists, and between the student and the scientists as a group. Identifying similarities and differences (also known as comparing and contrasting) is one of nine research-based strategies for increasing student achievement as described by Marzano, Pickering, and Pollock (2001). Second, students engage in repeated reading of the biographies. Repeated reading has been shown to improve fluency (Samuels 1979) and comprehension (Therrien 2004). Students use this strategy to extract multiple types of information from the texts used in the inquiry. More information on science process skills, identifying similarities and differences, and repeated reading can be found in Appendix 1 (p. 269).

OBJECTIVES

- Express ideas about scientists in drawings
- Name scientists and relate their accomplishments
- Identify similarities and differences among scientists and between scientists and themselves
- Identify scientific behaviors and habits of mind
- Practice the following science process skills: observing, communicating, classifying, and experimenting
- Write poetry

STANDARDS ALIGNMENT

National Science Education Standards
SCIENCE AS INQUIRY

- K–4, 5–8 Abilities Necessary to Do Scientific Inquiry

HISTORY AND NATURE OF SCIENCE

- K–4, 5–8 Science as a Human Endeavor
- 5–8 History of Science

Common Core State Standards for English Language Arts
INFORMATIONAL TEXT

- Grades 3–5 Key Ideas and Details
-

WRITING

- Grades 3–5 Text Types and Purposes

SPEAKING AND LISTENING

- Grades 3–5 Comprehension and Collaboration

For a detailed standards alignment, see Appendix 3 (p. 282).

TIME FRAME

- Fifteen 45-minute class periods

SCIENTIFIC BACKGROUND INFORMATION

John James Audubon took an interest in birds at a young age. He observed and drew birds and conducted the first bird-banding experiment in North America, in which he learned that birds return to the same nesting sites each year. He later produced *The Birds of America*, a text containing descriptions and original paintings of hundreds of birds (National Audubon Society n.d.).

William Beebe was interested in nature from childhood, leading him to take a job as the assistant curator of birds at the Bronx Zoo. Beebe later made several expeditions around the world to study and observe birds in their natural habitats. When he began diving in the ocean, he wanted a way to travel deeper to see the creatures of the ocean in their habitats. He and Otis Barton invented the bathysphere, a device that permitted deep-sea diving (The Official William Beebe Web Site n.d.).

As a child, Rachel Carson aspired to be a writer. She studied marine biology and later received a master's degree in zoology. She worked as a scientist and editor for the U.S. Fish and Wildlife Service, continuing to use her talents as a writer to introduce others to the beauty and mystery of the natural world. Her most famous work, *Silent Spring*, warned the population about the dangers of pesticides and changed policy about their use (Lear 1996).

Benjamin Banneker was born to freed black slaves in Maryland in 1731. He inherited the farm left to him by his grandparents, but continued to explore his other interests, including astronomy. Despite numerous rejections, in 1792 he published an almanac. Banneker also wrote to Secretary of State Thomas Jefferson with his almanac and a letter criticizing Jefferson for his support of slavery. Their correspondence was eventually published as well ("Benjamin Banneker Biography" n.d.).

Barnum Brown, paleontologist, was known as the greatest dinosaur hunter of them all. His team was the first to discover *Tyrannosaurus rex* skeletons. Many of his finds are on display at the American Museum of Natural History (American Museum of Natural History n.d.)

Julie Codispoti is a geologist and was the former assistant curator of the U.S. Polar Rock Repository at the Byrd Polar Research Center at Ohio State University in Columbus, Ohio.

Gregor Mendel had a keen interest in science and nature all his life. A monk and a high school science teacher, he wondered about how the traits that he observed in living organisms were inherited. He decided to perform careful experiments crossing pea plants with different traits and observing the traits present in the offspring. Mendel was the first to apply mathematics to describe the patterns of inheritance he observed in his experiments. Although his work was not accepted at the time, he would later become known as the Father of Genetics (Rhee n.d.).

Léon Foucault began his studies in medicine, but changed to physics. He was interested in questions about light and energy, taking the first picture of the Sun and measuring the speed of light more accurately than anyone who came before him. His most famous experiment was designing a pendulum that demonstrated that the Earth turned on its axis ("Jean Bernard Léon Foucault" 2004).

MISCONCEPTIONS

Misconceptions about scientists and their work have been studied at length, often through the use of the Draw-A-Scientist Test (DAST). Chambers first reported DAST results in 1983. Since then the test has been used in a multitude of studies in a large variety of contexts (Schibeci 2006). While there is some variation, the results typically reveal that children (and very often adults) hold a stereotypical view of scientists. And even though children often demonstrate an affinity for science, this view may stand in the way of children pursuing science studies or scientific careers (Archer et al. 2010). Some misconceptions about scientists are described in Table 6.1 (p. 48).

Sharkawy (2010) found that reading stories about scientists helped children develop a greater appreciation for scientists and their work. Encountering scientists as classroom guests, participating in citizen scientist projects such as Journey North (*www.learner.org/ jnorth*), and engaging in authentic scientific investigations are also helpful in addressing students' misconceptions about scientists and their work.

TABLE 6.1. COMMON MISCONCEPTIONS ABOUT SCIENTISTS

Common Misconception	Scientifically Accurate Concept
Science is dangerous; for example, it can cause explosions.	Great care and attention to safety considerations is common practice among scientists.
Scientists are Caucasian males. Scientists wear glasses, lab coats, and have funny hair. Scientists are middle-aged to old.	Science is practiced by a diverse group of individuals.
Science is hard.	Science is a way of thinking about natural phenomena that when first encountered may seem quite different from daily thinking.
Science is for "eggheads."	Science and science-related careers are practiced by a diverse group of individuals.
Scientists work alone.	Scientists cooperate and collaborate with one another and with nonscientists.
Scientists work in laboratories.	Scientists work in a variety of settings.
There is only one "scientific method."	There are many ways to conduct scientific investigations. The practice of science is better described by a set of processes than a single "scientific method."

TEXT SET

Barnum Brown: Dinosaur Hunter by David Sheldon (New York: Walker and Company, 2006); biography, Flesch-Kincaid reading level 6.5.

Barnum Brown had a gift for finding dinosaur bones, including the first *Tyrannosaurus Rex*. Learn more about this colorful paleontologist.

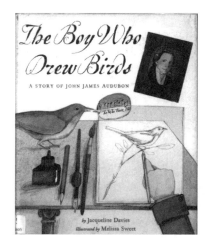

The Boy Who Drew Birds: A Story of John James Audubon by Jacqueline Davies (New York: Houghton Mifflin, 2004); biography, Flesch-Kincaid reading level 4.4.

John James Audubon loved birds from the time he was a little boy. Through his patient observations, he answered questions that had baffled scientists for years.

Come See the Earth Turn: The Story of Léon Foucault by Lori Mortensen (New York: Random House, 2010); biography, Flesch-Kincaid reading level 5.2.

In this beautifully illustrated book, learn about the man who proved that the Earth rotated on its axis.

Dear Benjamin Banneker by Andrea Davis Pinkney (San Diego, CA: Harcourt, 1994); biography, Flesch-Kincaid reading level 6.4.

Benjamin Banneker wanted to learn about the mysteries of the Sun, Moon, and stars. Along the way he created an almanac.

Gregor Mendel: The Friar Who Grew Peas by Cheryl Bardoe (New York: Harry N. Abrams, 2006); biography, Flesch-Kincaid reading level 6.0.

Gregor Mendel loved learning about nature, and he was the first to use mathematics and an experimental procedure to study and describe what he observed in nature. Learn more about the Father of Genetics in this well-written book.

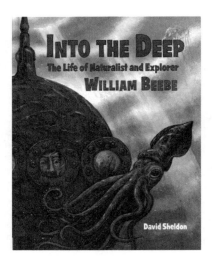

Into the Deep: The Life of Naturalist and Explorer William Beebe by David Sheldon (Watertown, MA: Charlesbridge, 2009); biography, Flesch-Kincaid reading level 6.2.

William Beebe traveled the world to observe and discover new species. When he began studying deep-sea creatures, he knew there was a better way to learn about them. So he and Otis Barton invented the bathysphere, a deep-sea diving vessel.

Rachel: The Story of Rachel Carson by Amy Ehrlich (San Diego, CA: Harcourt, 2003); biography, Flesch-Kincaid reading level 5.6.

Rachel Carson loved studying nature. But she also loved sharing what she learned with others. Learn how she used her talents to protect the environment.

Reader of the Rocks by Stephen Whitt (Columbus: Ohio State University, 2008) (Go directly to the electronic book for grades 4–5: *http://bit.ly/oO5H5o* and the illustrated book: *http://bit.ly/qCRjuP* [QR Code 1]); reference, Flesch-Kincaid reading level 5.7.

This text introduces students to the discipline of geology, and specifically polar geology. Students meet Julie Codispoti, assistant curator of the U.S. Polar Rock Repository, and learn how she overcame an initial dislike of science to become a geologist. Links provided are for the highest versions of the text. For modified versions at lower reading levels (1.9 and 3.5) along with grades 4–5, visit *http://bit.ly/uBwBUg* (QR Code 2).

QR Code 1   QR Code 2

S Is for Scientists: A Discovery Alphabet by Larry Verstraete (Ann Arbor, MI: Sleeping Bear Press, 2010); poetry, Flesch-Kincaid reading level 7.6.

"In this book/you will find/stories about scientists/with curious minds." So begins this rhyming alphabet book that pairs 26 scientific behaviors (adapt, experiment, hypothesize, and so on) with short biographical sketches of scientists. Teachers can use only the poetry for a shorter read-aloud, or they can use the longer biographies for a more detailed look at scientists.

MATERIALS

- Chart paper and markers
- String (kite string works well)
- Yardstick
- Magnifying glasses
- Student science journals
- Small objects for classification, such as rocks, feathers, leaves, fossils, or shells

- Objects to be used as pendulum bobs, such as washers or fishing weights of various masses
- Masking tape or other strong tape
- Metersticks
- Stopwatches
- Large sheets of white construction paper or poster board (for posters)
- Copies of supporting documents

SUPPORTING DOCUMENTS

- "Who Is a Scientist?" graphic organizer (p. 61)
- "Pendulum Investigation" worksheet (p. 62)

SAFETY CONSIDERATIONS

During the "Square Foot" activity, instruct students not to handle anything inside or outside of the area. During the "Pendulum Investigation," instruct students to stand clear of the swinging pendulum.

Scientists Like Me Inquiry Unit

ENGAGE

The engage phase begins the conversation around the characteristics and work of scientists.

1. Invite students to have a conversation about scientists. Do they know of any scientists? What do they know about them? What do they think scientists do? How did they learn about scientists? Record student answers on chart paper as a concept map, a web, or lists underneath each question.

2. Invite students to each draw a picture of a scientist and write a short paragraph explaining what they know about scientists using the "Who Is a Scientist?" graphic organizer. Allow students to share their pictures and explanations, and create a frequency table to display classroom trends. Collect and review student responses as a means of formative assessment to help you determine students' current understanding of scientists. Do they have a stereotypical view of scientists? What ideas (correct and incorrect) do they hold about scientists and their work? Keep student papers until the end of the explain phase.

3. Introduce the inquiry questions: *Who can become a scientist?* and *What do scientists do?* Post these questions in a prominent location in the classroom. Explain to students that they will listen to and read several biographies of scientists to find the answers to these questions.

Assess this phase: Student discussion, drawings, and paragraphs serve as formative assessment for this phase of the inquiry. They provide insight into students' current understanding of scientists and of the nature of science. If students hold limited or stereotypical views, there is no need to correct them at this time. Students will be exposed to a variety of scientists throughout the explore phase.

EXPLORE

This phase of the inquiry introduces students to pairs of scientists that exemplified the use of four scientific behaviors (science process skills): observing, communicating, classifying, and experimenting. Students identify similarities and differences between the scientists and then engage in firsthand experience with the process skills. By doing so, students are engaging in the process skills within the context of science learning, a best practice in science instruction. This phase is divided into five sections. The first four are each dedicated to the study of a pair of scientists and the development of one process skill. The final section asks students to identify similarities and differences across all eight scientists.

Part I. Science Process Skill: Observing
SCIENTISTS: JOHN JAMES AUDUBON AND WILLIAM BEEBE

1. Introduce the first two scientists to students by name. Explain that you are going to read two short biographies aloud. They should listen for scientific actions (behaviors) mentioned in each biography.

2. Read *The Boy Who Drew Birds* aloud, discussing the story as you read. Use questions as needed to ensure student comprehension of the story, but also encourage students to share observations and ideas without waiting for a specific question to be asked. A dialogic approach to reading in which students are able to freely offer their own ideas sets the stage for the remainder of the inquiry (Pappas et al. 2004; Wells 1999).

3. After finishing the book, draw a large T-chart on chart paper. Leave some blank space underneath the chart. Label one column with Audubon's name, and ask students to list scientific behaviors presented in the text.

4. Repeat the process with *Into the Deep*, adding behaviors to the other column of the T-chart under Beebe's name.

5. Ask students to identify similarities and differences between the two scientists, recording this information in the space underneath the T-chart. Focus student attention on the idea that both watched birds and animals in their natural habitat and recorded their observations through drawings and journals. Introduce and define the term *observation*, which is using the senses and tools to gather and record information about the natural world.

6. Take your students outside for a "Square Foot" observation activity. Before the activity, mark off square foot areas—each square 1 foot by 1 foot—with yarn or string. Ideally, have one square foot area per three or four students. Invite students to make careful observations of the area, using magnifying glasses if possible. Instruct students to record their observations using labeled drawings and sentences that explain what they observed.

7. Back in the classroom, discuss the experience with students. How did making their own observations give them insight into the work of Audubon and Beebe? What else have they learned about observation? Add students' responses to the T-chart. Keep this chart posted in the classroom through the remainder of the inquiry.

Part II. Science Process Skill: Communicating
SCIENTISTS: RACHEL CARSON AND BENJAMIN BANNEKER

1. Read *Rachel* aloud, discussing the story as you read. Again, encourage students to share ideas and thoughts while listening.

2. After finishing the book, draw a large T-chart on chart paper. Leave some blank space underneath the chart. Label one column with Carson's name, and ask students to list scientific behaviors described in the text.

3. Repeat the process with *Dear Benjamin Banneker*, adding behaviors to the other column of the T-chart under Banneker's name.

4. Ask students to identify similarities and differences between the two scientists, recording this information in the space underneath the T-chart. Focus student attention on the idea that both communicated their findings with others through writing.

5. Invite students to share what they observed by writing a letter to a friend or family member describing what they observed during the "Square Foot" activity. This does not need to be a polished final draft. Instead, the focus should be on the process of communicating scientific observations. Depending on the needs of your students, you may want to teach or review the parts of a letter before students can complete the assignment.

6. Once the letters are complete, discuss the experience with students. Did communicating their observations give them any insight into the work of Carson and Banneker? What else have they learned about communication? Add students' responses to the T-chart. Keep this chart posted in the classroom through the remainder of the inquiry.

Part III. Science Process Skill: Classifying
SCIENTISTS: BARNUM BROWN AND JULIE CODISPOTI

1. Read *Barnum Brown* aloud, discussing the story as you read. Again, create a dialogic read-aloud by inviting students to share their thinking, only using teacher-directed conversation if students need support in comprehending the text.

2. After finishing the book, draw a large T-chart on chart paper. Leave some blank space underneath the chart. Label one column with Brown's name, and ask students to list scientific behaviors described in the text.

3. Repeat the process with *Reader of the Rocks* by playing the electronic version of the text, adding behaviors to the other column of the T-chart under Codispoti's name.

4. Ask students to identify similarities and differences between the two scientists, recording this information in the space underneath the T-chart. Focus student attention on the idea that both classified objects.

5. Present students with a variety of natural objects to classify. Depending on your grade level and curriculum, you might choose to use rocks, feathers, leaves, fossils, or shells. Allow small groups of students to carefully examine the objects and devise their own system of classification. Students should present their classification scheme to the class. If you have additional time, consider having students draw a diagram that illustrates their classification scheme or write about (and illustrate) their criteria for classification.

6. Once the letters are complete, discuss the experience with students. Did classifying objects give them any insight into the work of Brown and Codispoti? What else have they learned about classification? Add students' responses to the T-chart. Keep this chart posted in the classroom through the remainder of the inquiry.

Part IV. Science Process Skill: Experimenting
SCIENTISTS: GREGOR MENDEL AND LÉON FOUCAULT

1. Read *Gregor Mendel* aloud, discussing the text and inviting students to share their thinking as you read.

2. After finishing the book, draw a large T-chart on chart paper. Leave some blank space underneath the chart. Label one column with Mendel's name, and ask students to list scientific behaviors described in the text.

3. Repeat the process with *Come See the Earth Turn*, adding behaviors to the other column of the T-chart under Foucault's name.

4. Ask students to identify similarities and differences between the two scientists, recording this information in the space underneath the T-chart. Focus student attention on the idea that both conducted experiments.

5. Explain to students that, like Foucault, they will conduct an experiment with a pendulum. (Students will generate the questions for the experiment now, but complete the actual investigation during the expand phase.) Demonstrate the pendulum by pulling it back and releasing it a few times.

6. Briefly talk about testable questions to help students understand the difference between questions that can and cannot be answered through investigation. Invite students to generate their own scientific question about the pendulum, using the following question stems:

   • What happens if ... ?

   • What would happen if you ... ?

   • What happens when ...?

Students should brainstorm possible scientific questions and record them on the "Pendulum Investigation" worksheet. Collect these questions to assess student understanding of scientific (testable) questions and determine whether students will be able to carry out experiments to answer their questions. If students are having trouble posing questions, plan to revisit this portion during the expand phase.

7. Once the questions are complete, discuss the experience with students. How did posing questions for experimentation give them any insight into the work of Mendel and Foucault? What else have they learned about experimentation? Add students' responses to the T-chart. Keep this chart posted in the classroom through the remainder of the inquiry.

Part V. Identifying Similarities and Differences

1. Ask students to review the four T-charts posted in the classroom. In pairs or small groups, students should draw a T-chart in their journals and use it to identify similarities and differences across all eight scientists. How were their behaviors the same? How were they different?

2. Read *S Is for Scientists* aloud, paying special attention to the behaviors highlighted through the rhyming text. Compare these with what students learned from the texts. Are the behaviors they've studied included? What other behaviors are mentioned? This read-aloud is more teacher directed than those previously included, because the purpose is to help focus student attention on a set of scientific behaviors.

Assess this phase: Formative assessment occurs throughout this first explore phase of the inquiry. As you read and discuss each pair of texts, monitor students' ability to identify scientific behaviors from the texts and identify similarities and differences between the two scientists. If students have difficulty, re-read portions of the texts and use guiding questions to help them focus on the relevant information. Formative assessment should also be used to determine student proficiency with the science process skills of observing, communicating, classifying, and generating testable questions (experimenting). We recommend using the "Science Process Skills Rubric" in Appendix 2 (p. 276) to assess student abilities in this area.

EXPLAIN

In the explain phase, students will share what they have learned about scientific behaviors to answer the question, *What do scientists do?*

Reread one page of *S Is for Scientists* aloud. This time, draw student attention to the layout and graphic elements of the page. Have small groups or pairs create a poster of one of the scientists with a short poem (or a paragraph) about a behavior he or she exemplified. These posters serve as one summative assessment task, so provide sufficient time for students to complete the task well. Display these posters in your classroom for the year, providing a constant reminder of important science process skills.

Assess this phase: The posters of scientists serve as one summative assessment task for this inquiry. The posters should demonstrate an understanding of the featured scientist as well as scientific behaviors. The "Science and Literacy Rubric" in Appendix 2 can be used to assess student performance, and the "Achievement Grading Standards" (also in Appendix 2) can translate this into a numerical grade. If students fall below 75%, return to the explore phase for additional work before returning to the writing prompt.

EXPLORE

In this second explore phase, students will return to the biographies in small-group settings. Repeated reading will allow them to gather evidence to answer the second question, *Who can become a scientist?*

1.   Divide students into pairs or small groups of three or four. Each group should have a copy of one of the eight books used in the first explore phase. Students will read the books in the small groups, paying attention to what the scientists were like. What personality traits did they display? What was their background?

2.   To record their observations, each group should trace the outline of one of their group members' bodies on chart paper. Inside this outline, students should write words and descriptive phrases about their scientist. Each group should share its poster with the class, explaining and providing evidence from the text as needed.

Assess this phase: Only formative assessment is needed in this phase. Circulate among the small groups as they read the biographies, providing support to students and groups that struggle to comprehend the text. Use guiding questions like "What does the text tell us about [scientist's name]?" and "What was [scientist's name]'s personality like, according to the text?" to help students focus on identifying descriptive words and phrases to record on their posters.

EXPLAIN

In this second explain phase, students will share and apply what they've learned about scientists' backgrounds and personalities. They will relate these findings to their own personalities and backgrounds.

1.   Each student should first complete a Venn diagram or T-chart comparing themselves with the scientists. How are they alike? How are they different? At the bottom of the page, have students complete the sentence stem, "I am a scientist because ..."

2.   Have students draw another picture of a scientist. Have students compare these drawings with the ones they did at the start of the inquiry. What have they learned? Have students write a short paragraph explaining how their understanding of scientists has changed as a result of the inquiry.

Assess this phase: Students' pictures and paragraphs serve as the second summative assessment task for the inquiry. They should reveal a more sophisticated conception of scientists and the nature of science. The "Science and Literacy Rubric" in Appendix 2 (p. 276) can be used to assess student performance, and the "Achievement Grading Standards" (also in Appendix 2 [p. 277]) can translate this into a numerical grade. If students fall below 75%, return to the explore phase for additional work before returning to the writing prompt.

EXPAND

In this phase, students will apply what they've learned about scientists and the science process skills they have used by investigating a variable related to the pendulum introduced earlier.

1. Have students return to the questions they posed earlier. You may need to spend more time helping students generate testable questions, depending on their success with the earlier activity. If so, consider modeling a few questions using the question stems. If students really struggle, you can generate a list of questions as a class and allow students to select one from the list.

2. The "Pendulum Investigation" worksheet will guide students through the process of planning and conducting their investigations. You may choose to work through the worksheet as a class if students do not have much experience with conducting investigations independently. If students are more comfortable with this process, we recommend letting them work through the process in pairs or small groups, providing support and assistance as needed.

3. In addition to completing the worksheet, students should prepare a poster sharing their investigation and the results. Posters can be created on large sheets of construction paper or on poster board. Posters should include the question studied, the materials and procedure, data, and a conclusion.

4. Students will share their posters with their peers in a scientific symposium. This can be as simple as having half of the students set up their posters around the classroom and having the other half circulate around the room to view the posters and ask questions of the students (also known as a gallery walk).

Assess this phase: This is a time to assess science process skills involved in the pendulum investigation. Assessment should be focused not on a correct answer but on the process skills themselves. We recommend using the "Science Process Skills Rubric" in Appendix 2 to assess student abilities in this area.

REFERENCES

American Museum of Natural History. n.d. Personalities in paleontology: Barnum Brown. *www.amnh. org/exhibitions/permanent/fossilhalls/personalities/bios/brown.php*

Archer, L., J. DeWitt, J. Osborne, J. Dillon, B. Willis, and B. Wong. 2010. "Doing" science versus "being" a scientist: Examining 10/11-year-old schoolchildren's constructions of science through the lens of identity. *Science Education* 94 (4): 617–639.

Benjamin Banneker biography. In *Encyclopedia of world biography.* n.d. *www.notablebiographies. com/Ba-Be/Banneker-Benjamin.html*

Chambers, D. 1983. Stereotypic images of the scientist: The Draw-A-Scientist Test. *Science Education* (67) 2: 255–265.

Jean Bernard Léon Foucault. 2004. In *Encyclopedia of World Biography. www.encyclopedia.com/ doc/1G2-3404702238.html*

Lear, L. 1996. The life and legacy of Rachel Carson. *www.rachelcarson.org*

Marzano, R. J., D. J. Pickering, and J. E. Pollock. 2001. *Classroom instruction that works: Research-based strategies for increasing student achievement.* Alexandria, VA: Association for Supervision and Curriculum Development.

National Audubon Society. n.d. John James Audubon. *www.audubon.org/john-james-audubon*

The official William Beebe website. n.d. *https://sites.google.com/site/cwilliambeebe/Home*

Padilla, M. J. 1990. The science process skills. *Research Matters—to the Science Teacher,* No. 9004. *www.narst.org/publications/research/skill.cfm*

Pappas, C. C., M. M. Varelas, A. Barry, and A. Rife. 2004. Promoting dialogic inquiry in information book read-alouds: Young urban children's ways of making sense in science. In *Crossing borders in literacy and science instruction,* ed. E. W. Saul, 161–189. Newark, DE: International Reading Association.

Rhee, S. Y. n.d. Gregor Mendel (1822–1884). *www.accessexcellence.org/RC/AB/BC/Gregor_Mendel.php*

Samuels, S. J. January, 1979. The method of repeated readings. *The Reading Teacher* 32 (4): 403–408.

Schibeci, R. 2006. Student images of scientists: What are they? Do they matter? *Teaching Science—the Journal of the Australian Science Teachers Association* 52 (2): 12–16.

Sharkawy, A. 2010. Moving beyond the lone scientist: Helping 1st-grade students appreciate the social context of scientific work using stories about scientists. *Journal of Elementary Science Education* 21 (1): 67–78.

Therrien, W. J. 2004. Fluency and comprehension gains as a result of repeated reading: A meta-analysis. *Remedial and Special Education* 25 (4): 252–261.

Wells, G. 1999. *Dialogic inquiry: Towards a sociocultural practice and theory of education.* Cambridge, England: Cambridge University Press.

SCIENTISTS LIKE ME

Name_____ Date_____

WHO IS A SCIENTIST?

Draw a picture of a scientist.

Write a paragraph explaining what you know about scientists.

_____

_____

_____

_____

_____

Name_____ Date_____

PENDULUM INVESTIGATION

Questions

What will you investigate?

What happens if _____

What would happen if you _____

What happens when _____

Other questions I have:

_____

_____

Circle the question you are going to test in your investigation.

Materials

What will you need for your investigation?

- _____
- _____
- _____

Procedure

What steps will you follow?

1. _____
2. _____
3. _____
4. _____

Name_____ Date_____

PENDULUM INVESTIGATION *(Continued)*

Data

What information will I collect? What will I measure?

- _____

- _____

Data Table

Record your information here.

Variable (What changed?)	Trial 1	Trial 2	Trial 3

Conclusion

What did I learn? How does my evidence support my conclusion?

_____

_____

_____

Chapter 7
Measuring Pennies and More

OVERVIEW

Through this inquiry, students work to answer the question, *How does measurement help us know about the world?* The focus of the inquiry is what measurement can tell us. A scavenger hunt helps students learn to use measurement to describe everyday items, and a comparison activity helps them learn to select appropriate measurement tools. During the course of the inquiry, students learn about ordinal measurement scales and conduct an investigation in which measurement is critical.

As part of the inquiry, students write an experimental protocol and develop an ordinal scale for a set of items or a natural phenomenon. In doing this, the inquiry fosters students' abilities to use descriptive language and write procedural information.

OBJECTIVES

- Explain the importance of measurement in science
    - » Description
    - » Clarity
    - » Reproducibility
    - » Communication
- Practice using a variety of measuring tools
- Conduct an investigation in which measurement is a critical component
- Devise an ordinal measurement scale
- Write an interview for a newscast and conduct the interview for an audience

STANDARDS ALIGNMENT

National Science Education Standards
SCIENCE AS INQUIRY

- K–4 Abilities Necessary to Do Scientific Inquiry

Common Core State Standards for English Language Arts
WRITING

- Grades 3–5 Text Types and Purposes

Common Core State Standards for Mathematics
MEASUREMENT AND DATA

- Grade 3 Represent and Interpret Data

For a detailed standards alignment, see Appendix 3 (p. 282).

TIME FRAME

- Nine 45-minute class periods

SCIENTIFIC BACKGROUND INFORMATION

Measurement is an integral part of science. It helps remove ambiguity, provides a common language, and helps scientists communicate with clarity what they have learned. Measurement allows us to describe the natural world in greater detail. Our capacity to make accurate measurements is limited by the measurement tools at our disposal. Accuracy in measurement refers to how close the measurement is to the actual quantity of the thing we are measuring. Selecting the proper tools is important and can help us achieve greater accuracy in our measurements.

Generally, when we think of measurement, we think of using a calibrated tool to determine an object's dimensions. We then record that measurement with numbers and a unit. Another type of measurement, *ordinal measurement*, is more descriptive. When using ordinal measurements, we rank or order the characteristics in some logical way. We then assign a number or letter to each ranking. In ordinal measurement scales, the degrees separating the numbers or letters may vary. For example, rating movies with 1 to 5 stars could be considered an ordinal scale. We would apply 5 stars to a movie that was excellent and 4 stars to a movie that was pretty good but not quite as good as a 5-star movie. The degree of difference between a 5-star and a 4-star movie may not be exactly the same as the degree of difference between a 4-star and a 3-star movie. But there are clearly distinguishable differences between each of the star levels. Examples of ordinal scales used in science include the Beaufort scale for wind intensity, the Mohs scale of hardness for minerals, and the Fujita-Pearson Tornado Intensity Scale.

MISCONCEPTIONS

Measurement is an extremely important process in science. Misconceptions about measurement can be addressed in a variety of ways in both mathematics and science classrooms. The misconceptions in Table 7.1 come from "Children's Misconceptions about Science" (Operation Physics 1998).

TABLE 7.1. COMMON MISCONCEPTIONS ABOUT MEASUREMENT

Common Misconception	Scientifically Accurate Concept
Any quantity can be measured as accurately as you want.	The accuracy to which a quantity can be measured depends on the calibration of the instrument. No measurement is 100% accurate.
The metric system is more accurate than other systems.	Accuracy is how close a measurement is to the true value of the measurement. It is not dependent on the measurement system used.
The English system is easier to use than the metric system.	The metric system is based on powers of 10. It is the generally accepted system of measurement used in science. The perception of ease of use may be related to familiarity.
You can only measure to the smallest unit shown on a measuring device.	Measurement includes an estimation of the value beyond to the smallest increments on the device. For example, when using a ruler with centimeters as the smallest marked unit, the measurement is estimated to the nearest millimeter.
You should start at the end of the measuring device when measuring distance.	The ends of measuring devices such as rulers are often damaged. To ensure the greatest degree of accuracy, measurement typically begins with the first intact marking. Measurement could begin with any intact mark on the ruler.
Some objects cannot be measured because of their size or inaccessibility.	Measurements can be made indirectly.
Mass and weight are the same and they are equal at all times.	Mass is the amount of matter in an object. In the metric system mass is measured in grams. Mass is a fundamental unit. Weight is a derived unit that is the product of an object's mass and the force of gravity. Barring alterations an object's mass remains constant but its weight changes as the force of gravity changes.
Mass is a quantity you get by weighing an object.	Weighing is measured with a scale; mass is measured with a balance.
Mass and volume are the same.	Mass is the amount of matter in an object, volume is how much space it takes up.

Continued on next page

Continued from previous page

Common Misconception	Scientifically Accurate Concept
The only way to measure time is with a clock or watch.	Time can be measured in a variety of ways. Historically time has been measured with sundials, water clocks, and hourglasses to name a few. Sidereal time and universal time are examples of time based on astronomy.
You cannot measure the volumes of some objects because they do not have "regular" lengths, widths, or heights.	The volume of irregularly shaped objects can be determined in a variety of ways. Water displacement is a simple way to measure the volume of such an object.

As with other misconceptions, misconceptions about measurement are sometimes the result of differences between scientific and daily language. For example, the term *weight* is commonly used interchangeably with the term *mass,* but from a scientific perspective these are related but fundamentally different terms. Using the correct scientific term and the correct tools in class is one step toward correcting measurement misconceptions. Another is modeling correct use of measuring devices. Connecting children's literature to relevant hands-on activities may also help alleviate some measurement misconceptions (Tucker, Boggan, and Harper 2010).

TEXT SET

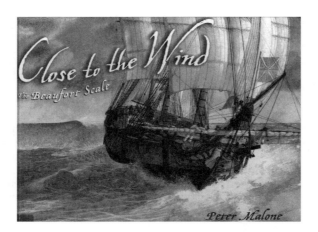

Close to the Wind: The Beaufort Scale by Peter Malone (New York: Putnam, 2007); narrative expository, Flesch-Kincaid reading level 6.8.

In this fictional account set in 1805, a British man-of-war sails across the Atlantic into a fierce storm. On left-hand pages, a loglike entry describes one of the Beaufort scale values, a fragment of correspondence from a 12-year-old boy recounts life aboard ship, and a paragraph of text explains interesting facts. Paintings on the right-hand pages capture moments in the voyage.

The following four books by Thomas K. Adamson and Heather Adamson (Mankato, MN: Capstone Press, 2011) are part of the same series, all written in a simple and straightforward manner. They share a common format that includes large, brightly colored photographs showing measurement tools and children making measurements. Customary and metric units of measurement are used throughout, and each book ends with a list of "cool measuring facts."

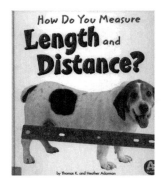 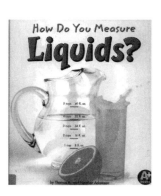  

How Do You Measure Length and Distance?; explanation, Flesch-Kincaid reading level 2.8.

How Do You Measure Liquids?; explanation, Flesch-Kincaid reading level 3.2.

How Do You Measure Time?; explanation, Flesch-Kincaid reading level 2.8.

How Do You Measure Weight?; explanation, Flesch-Kincaid reading level 3.5.

How Tall, How Short, How Far Away by David A. Adler (New York: Holiday House, 1999); reference, Flesch-Kincaid reading level 4.4.

How did people measure things in ancient Egypt? And what method did they use in ancient Rome? Students learn about these two ancient measurement systems as well as the metric and customary systems. Fun challenges include using the early Egyptian techniques and choosing the most appropriate units of measure.

Measuring Penny by Loreen Leedy (New York: Henry Holt, 1997); narrative expository, Flesch-Kincaid reading level 3.2.

Lisa measures her dog, Penny, from nose length and paw size to height, weight, and the time it takes to reach the food dish. And she does this using standard units (e.g., inches and centimeters) and nonstandard units (e.g., dog biscuits and cotton swabs). The illustrations are child (and dog) friendly.

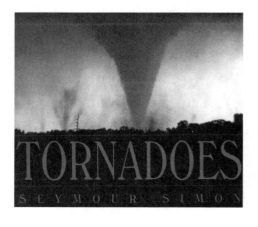

Tornadoes by Seymour Simon (New York: Morrow Junior Books, 1999); reference, Flesch-Kincaid reading level 6.4.

This book includes a detailed description of the Fujita-Pearson Tornado Intensity Scale. It also includes full-page photographs of tornadoes and their aftermath.

MATERIALS

- A variety of rulers, including metersticks and yardsticks
- A variety of scales, such as spring scale, bathroom scale, kitchen scale, and electronic scales
- A variety of timing devices, such as stopwatch, clock with second hand, clock without second hand, and hourglass
- Graduated cylinders, teaspoons, and measuring cups
- String (kite string works well)
- A variety of objects to be used as pendulum bobs; washers or fishing weights of various masses work well

- Masking tape or similar strong tape
- Scissors
- Pipette or eyedroppers
- Pennies
- Text set, including multiple copies of How Do You Measure Length And Distance? How Do You Measure Liquids? How Do You Measure Time? and How Do You Measure Weight? if possible.
- A variety of items ranging from very small to large that can be measured and weighed:
  - » Length: penny, paper clip, grain of rice, string, desk
  - » Weight: penny, grain of rice, textbook, brick
  - » Volume: ¼ tsp., 2 c., 50 ml water samples with different-color water
  - » Time: time it takes to walk across the room, time for ice to melt, time for paper to drop from hand to floor
- A variety of unusual items that can be measured by teams of students (e.g., a block of paraffin, a multicolored plastic cube)
- Riddle cards (teacher prepared, see "Measurement Scavenger Hunt" advance preparation, p. 72)
- Hint cards (teacher prepared, see "Measurement Scavenger Hunt" advance preparation, p. 73)
- Event cards (teacher prepared, see "Part I: Measuring Pennies and More" advance preparation, p. 75)
- Document camera (optional)

SUPPORTING DOCUMENTS
- "Measuring Pennies and More"
- "Seed Discussion Graphic Organizer"
- "Measuring Pennies and More: Pendulum Investigation"
- "In the News Interview Planner"

SAFETY CONSIDERATIONS
During the pendulum investigation, instruct students to stand clear of the swinging pendulum.

Measuring Pennies and More Inquiry Unit

ENGAGE

In the engage phase students begin thinking about measurement as a means for clearly describing and communicating observations about objects. Direct instruction on how to use a ruler and a scale may be needed before this activity if students do not have experience using rulers and a scale. Alternatively, you could observe teams as they measure and weigh items and provide assistance as needed.

Measurement Scavenger Hunt

Advance preparation: Prepare riddle cards (see Figure 7.1) that describe various items in your classroom by their physical dimensions and weight. Write each riddle on a separate index card.

FIGURE 7.1. SAMPLE RIDDLE CARD

RIDDLE CARD _____

The item is 15 cm long, 5 cm wide, and 4 cm high.

The item's weight is 60 g.

When not in use the item is perched about 1 meter above the ground.

What is it?

Also prepare hint cards (see Figure 7.2, p. 73) that describe the physical characteristics of each item on the riddle cards. Select items for the scavenger hunt that have to be measured and weighed to differentiate them from other similar items (e.g., a whiteboard eraser that is slightly smaller than others in the room). Students will begin the scavenger hunt with the riddle cards. They will only be given the hint cards if they need additional information to locate the item described on the riddle card.

FIGURE 7.2. SAMPLE HINT CARD

HINT CARD _____

The item is soft and fuzzy on one side and hard on the other.

The soft fuzzy side is dark in color.

The item is used to clean marks off of a white surface.

What is it?

1. Divide the students into teams of two. If possible have a scale available for each team; otherwise have a weighing station with scales available for teams to use as needed.

2. Give each team a riddle card and a ruler. Instruct the teams to find the object described on their card. Set a time limit for finding the items. Draw attention to the fact that you are setting the stopwatch, timer, or whatever timing device you are using. Later you will ask students what else was measured during the activity. Calling attention to setting the time now will help them remember later. Circulate through the room as students are looking for their items. If teams are having a hard time getting started, provide individual prompts that send them in the right direction (e.g., "Have you looked near the front of the room?"). When students ask, or when it is clear a team is having trouble finding its item, give the team the related hint card.

3. After the teams have located their items, use the following guiding questions to lead a class discussion about the experience.
   - What information was most helpful to you when you were looking for the object?
   - What information was least helpful?
   - How did the measurements help you distinguish the scavenger hunt item from other similar items?
   - What characteristics of the items did we measure?
   - Have we measured anything else today? (Students should mention time.)

4. As the discussion is drawing to a close, summarize what was measured in the activity (length, weight, time). Ask students to share when they have measured or observed others measuring things in the past.

5. Read *Measuring Penny* to the class. As you read, students will record the characteristics Lisa measured and the units she used. Allow students to choose the method they will use to record the measurements.

6. Make a list of the characteristics and units students have recorded. Organize the list as a two-column graphic organizer, with one column labeled "Characteristic" and the other "Unit." Marvel with your students about the number of Penny's characteristics that Lisa measured. Ask, "Why would Lisa care about Penny's measurements?"

7. Instruct students to select an item at home to measure in as many ways as they can. They will share their results the next day. Inform them that they are welcome to use standard or nonstandard measures, or both. As a class, determine the minimum number of characteristics each student should measure. Students should construct another two-column graphic organizer to record their measurements. One column should be labeled "Characteristic" and the other "Measurement." Remind students that the measurement must include the unit.

Assess this phase: Formative assessment is used at this point in the inquiry to check student progress and help you determine if instructional modifications are needed. Observe students as they participate in the "Measurement Scavenger Hunt" and provide assistance as needed. Students' ability to weigh and measure items will help determine how much support to provide during the explore phase.

EXPLORE

In the first part of the explore phase, students compare different measuring tools and select the best ones for making large and small measurements. In the second, they design and conduct a pendulum investigation that requires careful measurement.

Part I: Measuring Pennies and More

Advance preparation: Gather tools to measure weight, length, time, and volume. The tools for each measurement type should vary in their calibration; for example, instruments for measuring volume should vary from a ¼ tsp. measuring spoon to a 2 c. measuring cup. Be sure to include both customary and metric measuring tools. Include items from comparatively large to very small (e.g., from a grain of rice to a brick for weight) for students to measure. Prepare event cards (see Figure 7.3, p. 75) for use in the time station described below. Each event card should describe an action the students will measure, such as time it takes to walk across the room, draw a line on the chalkboard from one end to the other, or write their name 10 times. Before class set up length, weight, volume, and time stations.

FIGURE 7.3. SAMPLE EVENT CARD

EVENT CARD 1

Measure the amount of time it takes to walk across the room.

1. After the *Measuring Penny* assignment is completed, discuss what students found and how they chose to measure the item they selected. Guide students toward recognizing that base units of measurement (e.g., length, weight, time) are divided into smaller and smaller units or are combined into larger and larger units. The following guiding questions will be helpful:

   - What characteristic of the item was the easiest to measure? The most difficult?
   - What was the biggest measurement taken? The smallest?
   - How did measuring the item help you know more about it?
   - What tools did you use?
   - What did you do when the object did not line up with the markings on the tool?

2. The next set of guiding questions is designed to lead students toward thinking about the calibration of measuring tools. Before you begin this set of questions, draw on the board an oversized ruler with markings that divide each unit into quarters, as shown in Figure 7.4, or use a document camera and place a ruler on the document camera for the demonstration.

FIGURE 7.4. OVERSIZED RULER

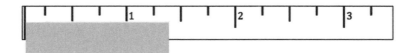

3. Demonstrate that items being measured often do not line up with the marks, making it difficult to get an exact measurement. While you are holding an item that does not line up with the marks, point out that the there is a gap between the edge of the object and the

marking on the ruler. Discuss with students how they handled similar situations. Ask,

- What did you do when that happened? (Students will likely respond that they guessed or estimated. This would be a great time to talk about estimation.)
- How can you make measuring very small or very large things more exact?

4. Read *How Tall, How Short, How Far Away?* pausing to point out how base units are divided into smaller units or combined into larger units. After reading the book, tell students that they are going to determine which measuring tools are best for measuring a variety of items.

5. Organize students into cooperative groups of three to four students. Distribute a penny to each group and a "Measuring Pennies and More" data collection sheet to each student. Review the procedure on the "Measuring Pennies and More" form. Explain to students that at the time station they will be selecting event cards rather than items. Each student in the group should have a chance to work with the various measuring tools.

6. Using the following questions, lead a class discussion about the experience after all groups have finished.

- How did the different measuring tools at each station compare with one another?
- What did you notice about using different measuring tools to measure the same item or event?
- Which tools were the best for measuring the smallest items at each station? The largest items at each station?

7. Reorganize the students into groups of five. Working in these small groups, students will reinforce what they have learned about measurement by reading *How Do You Measure Length and Distance? How Do You Measure Liquids? How Do You Measure Time?* and *How Do You Measure Weight?* Each group will select and read only one of these books. Students will then use a Seed Discussion strategy to facilitate their small-group discussions. Details about this strategy are provided below.

8. Lead a class discussion in which each group shares what they have learned with the rest of the class. Paired reading, an alternative grouping approach that can be used with the Seed Discussion, is described below. Select the option that best fits the needs of your students.

A Seed Discussion is a strategy through which students respond in writing to a predetermined set of prompts as they are reading. The students' responses to the prompts serve as "seeds" for the discussion. During the discussion, group members individually share their seeds. The remaining group members comment on the seed of a member before the next group member shares. Group members may be assigned roles to help keep the group discussions orderly and productive. In this modified version of a Seed Discussion, one group member will read the assigned book aloud to the remaining members (Messmer 2009).

1. Divide the class into groups of five. Each group of five will work with a different book. Assign the following roles to the group members:

   - Reader: Reads the book aloud to the group.

   - Leader: Leads the discussion by calling on group members to share their seeds, making sure to call on all group members.

   - Manager: Makes sure all group members have the necessary materials.

   - Timekeeper: Makes sure that the group stays on track to complete sharing within the allotted time.

   - Communicator: Lets the teacher know when the group is finished and ready to move on to the class discussion.

2. Give each student a "Seed Discussion Graphic Organizer." Review and model the procedure. As the groups read and discuss the books, circulate around the room providing guidance as needed. Monitor group discussions, noting any areas that may need review before moving forward. When the groups have finished, lead a discussion in which each group shares some of their seeds and highlights of the group's discussion. Make the books available for students to read at their discretion.

Paired reading alternative: Organize students into pairs of mixed-ability readers. Pair high-achievement readers with midlevel readers, and pair midlevel readers with low-achievement readers. The pair takes turns reading to one another. The length of the passage to be read can vary from a single sentence to the entire page. Typically the more accomplished reader begins, followed by the second reader. While one student is reading, the other reads along asking questions and offering support as needed. As they are reading, the pair can add seeds to their "Seed Discussion Graphic Organizer." When all pairs have finished their books, lead a class discussion in which students share their seeds.

Part II: Getting Into the Swing With Measurement

The objective of this activity is to use newly developed measurement skills in a student-directed investigation.

1. Prior to class suspend a pendulum (made by tying a length of string to the pendulum bob) in a location where everyone in the class can easily observe it.

2. Begin by congratulating the class for learning so much about measurement. Tell students that they are going to use what they have learned about measurement to answer the question, *How does measurement help us know about the world?*

3. Demonstrate the pendulum by pulling it back and releasing it. Repeat several times, changing the angle of release each time. Ask students to share their observations as the pendulum swings. After a few minutes ask, "What characteristics of a pendulum can we measure?" (Student responses should include length of string, weight of the bob, and time of swing.). After the students have identified the measurable characteristics, brainstorm some testable questions. Possibilities include:

   • How can we make the pendulum swing faster? Slower?

   • What happens when we change the length of the string? The weight of the bob?

4. Inform students that they will design a pendulum investigation in which they change one variable. They will either change the mass and keep the length of the string the same, or they will change the length of the string and keep the mass the same. It is up to the students to determine the procedure they will follow and the materials they will use. They must carefully measure the variables and count the number of swings in a set amount of time.

5. Divide the students into small groups. Provide each group with a "Measuring Pennies and More: Pendulum Investigation" data collection sheet. Instruct them to select the variable they will keep the same and the one they will change. Then each group should decide how they will conduct their investigation. Remind them that they will need to select their materials, including their measuring tools, and include a materials list in their procedure. Before beginning their investigation, they should get the teacher's approval for their procedure. Each group will share their results with the class.

6. As a class, discuss the results focusing on the role of measurement in the investigation. Emphasize the importance of measurement. These guiding questions will help:

   • Which variable made a difference in the speed of the pendulum's swing? (Student responses should note that the mass doesn't matter, but the length of the string makes a difference.)

   • If you want a pendulum to swing slower what would you do? Faster?

   • How did making careful measurements help you in your investigation?

   • How easy or difficult would it be to repeat another group's procedure if you didn't know the measurements they used?

   • How can describing things with measurements help us communicate what we know?

Assess this phase: Formative assessment is used in both parts of this phase. During Part I, observe students as they work with the variety of measuring tools and provide support as needed. Observe small groups as they engage in the Seed Discussions to ensure that all students are participating. Use the writing prompts on the "Seed Discussion Graphic

Organizer" to prompt students who are reluctant to participate or claim that they don't have anything to share. During Part II, observe students as they plan and conduct their investigations. Use guiding questions to help students plan their investigation and focus on relevant data. You might also choose to assess students' proficiency with the "Science Process Skills Rubric" in Appendix 2 (p. 277).

EXPLAIN

In the explain phase students share their expertise about measurement.

In the News

Advance preparation: Bring some unusual items in that students can measure and describe for their newscast. The items should be things the students are not likely to have previously encountered. Be sure to select items that students can successfully measure with available measuring tools. To enhance the sense of discovering a new item, place each item in a separate opaque bag or box, wrap them in newspaper or gift-wrap, or in some other way package them so students cannot identify the items. Then position the packages in various locations around the room where students will quickly "discover" them.

Begin by pairing students into teams and inviting each team to locate a package somewhere in the classroom. Instruct teams to wait until everyone has found a package before opening theirs. When all teams have found a package say, "You have discovered an unusual item. You want to learn everything you can about it and then share what you have learned with others. Open your packages to see what you have discovered."

1. Ask the class, "How would measuring the item help you learn more about it? How would you use the measurements to tell others about the discovery and the unusual item?"

2. Teams will now play the roles of science news reporter and scientist. Using the "In the News Interview Planner" they will write interview questions and responses for a local newscast in which they explain the unusual item the scientist just discovered. They will describe the discovery using at least three of the measurements they have studied. The interview planner should include questions about the discovery and answers that are rich descriptions that include measurements. The measurements must include units.

3. Each pair of students will then share their newscast with the class.

4. Ask students to write a response to the following prompt, *How does measurement help us know about the world?*

Assess this phase: Student-created newscasts serve as one form of summative assessment, and written responses to the writing prompt serve as a second. Both should demonstrate an understanding of the importance of measurement in science (description, clarity, reproducibility, and communication). The "Science and Literacy Rubric" in Appendix 2 can be used to assess student performance, and the "Achievement Grading Standards" (also in Appendix 2) can translate this into a numerical grade. If students fall below 75%, return to the explore phase for additional work before proceeding.

EXPAND

1. Explain to students that not everything is measured with rulers and scales. Some things are measured in other ways. For instance, T-shirt sizes include small, medium, and large. Shoes are measured by size, too, but shoe sizes use numbers. Some things are measured by comparing them with other things.

2. Read *Close to the Wind* and pages 15–19 of *Tornadoes*. Follow with these questions:
   - How does this type of measurement compare with the measurements we studied earlier?
   - How is this type of measurement scale useful?
   - What sorts of things in the school could be measured with a comparative measurement scale?

3. Brainstorm a list, select one characteristic, and work as a class to develop a measurement scale for the characteristic. Possible characteristics are the length of the lunch line, the loudness of the PA system, or the amount of artwork on the hallway walls. For example, a measurement scale for the loudness of the PA system might be based on a physical response to the beginning of the announcements: PA-1, whisper—when the announcements begin you must strain to hear them; PA-2, comfort—when the announcements begin you want to stop what you are doing and listen; and PA-3, scary—when the announcements begin they are so loud that they make you jump in your seat.

4. After the class has developed a comparative measurement scale, have students independently develop one for some characteristic of the classroom, such as the dustiness of erasers or noise level in the classroom.

Assess this phase: Ordinal measurement scales serve as the final summative assessment task for this inquiry. The "Science and Literacy Rubric" in Appendix 2 (p. 276) can be used to assess student performance, and the "Achievement Grading Standards" (also in Appendix 2) can translate this into a numerical grade. If students fall below 75%, return to the explore phase for additional work before returning to the writing prompt.

REFERENCES

Messmer, A. M. 2009. Seed discussions. *http://literacy.purduecal.edu/STUDENT/ammessme/Seed.html*

Operation Physics. 1998. Children's misconceptions about science. American Institute of Physics. *www.eskimo.com/~billb/miscon/opphys.html*

Tucker, C., M. Boggan, and S. Harper. 2010. Using children's literature to teach measurement. *Reading Improvement* 47 (3): 154–161.

Name_____ Date_____

MEASURING PENNIES AND MORE

1. Select three items from one of the measuring stations. Select three different measuring tools from the same station.
2. Return to your team's work area and measure each item with each of the measuring tools.
3. Record the item measured in the first column. In the following columns, record the measuring tool and the measurement. After measuring the item with three different tools, underline the measurement that is closest to the true measurement of the item.
4. Work with one set of materials at a time. Return them when you are finished.
5. Repeat this process until you have measured three objects from each station.

Length (Height)

Item	Measurement 1	Measurement 2	Measurement 3
Grain of rice	12 in. ruler 1/8 in.	15 cm ruler 2 mm	yardstick ¼ in.
Penny			

Name_____ Date_____

Weight

Item	Measurement 1	Measurement 2	Measurement 3
Building block	Kitchen scale 1 oz.	Electronic scale 32 g	Bathroom scale 0 lbs.
Penny			

Volume

Item	Measurement 1	Measurement 2	Measurement 3
Small glass of green water	1 c. measuring cup ¼ c.	Graduated cylinder 45 ml	Teaspoon 11½ tsp.

Name_____ Date_____

Time

Event	Measurement 1	Measurement 2	Measurement 3
Time to walk across the room	Stopwatch 11.7 seconds	Hourglass ¼ of the sand	Clock 15 seconds

What did you notice about using different measuring tools to measure the same item or event?

_____

_____

_____

_____

Name_____ Date_____

SEED DISCUSSION GRAPHIC ORGANIZER

Things that confirmed what I already know:	**Things that surprised me:**
Things that remind me of something else:	**Things I did not understand:**

Name_____ Date_____

MEASURING PENNIES AND MORE: PENDULUM INVESTIGATION

Variables

The variable we will keep the same is _____.

(weight of the bob or length of string)

The variable we will change is _____.

(weight of the bob or length of string)

Question

How will the rate of the pendulum's swing change if the <u>(weight of the bob or length of string)</u> is kept the same but the <u>(weight of the bob or length of string)</u> is changed? _____

_____

_____

_____

_____

Materials

What will you need for the investigation?

- _____

- _____

- _____

MEASURING PENNIES AND MORE

Name_____ Date_____

Procedure

What steps will you take?

1. _____

2. _____

3. _____

Data

What measurements will you take?

Number of swings in _____ seconds (15 or 30 seconds)

Record information for the variable you are changing.	
Length of string Short: Medium: Long:	**Weight of bob** Light: Medium: Heavy:

Data Table

Variable	Trial 1	Trial 2	Trial 3
Length or weight	5 swings	7 swings	6 swings

Name_____ Date_____

Conclusion

When the (<u>weight of the bob or length of string</u>) of the pendulum is changed and the (<u>weight of the bob or length of string</u>) is kept the same, the number of swings in (<u>15 or 30</u>) seconds (<u>changes, stays the same</u>).

Using the measurements you took during the investigation, describe what happened as you changed one variable.

Example: When the string was 30 cm long, the pendulum swung back and forth 7 times in 30 seconds. When the string was 60 cm long, the pendulum swung back and forth 10 times in 30 seconds. When the string was 90 cm long, the pendulum swung back and forth 14 times in 30 seconds.

_____

_____

_____

_____

_____

_____

Name_____ Date_____

IN THE NEWS INTERVIEW PLANNER

Opening: How will you introduce the scientist and the discovery to your viewers?

Now write three questions that the scientist can use measurements to answer.

Example question: What was the most amazing thing about your discovery?
Example answer: The most amazing thing was the weight of the item. It weighed only 1 g! This is the smallest item of this kind ever found!

Question 1	Question 2	Question 3

Name_____ Date_____

Answer _____	**Answer** _____	**Answer** _____

Conclusion: How will you summarize the story for your viewers?

Chapter 8
Minds-on Matter: Phase Changes and Physical Properties

OVERVIEW

In this inquiry, students work to answer the question, *How do the physical properties of a material change during phase changes?* Students observe and measure physical properties of solids, liquids, and gases. They also measure the changes in temperature during melting, freezing, and boiling. Students are introduced to glaciers and the study of ice cores, and they apply what they've learned to design containers that keep an ice cube from melting.

Students learn about cause-and-effect text structure and identify words that signal cause-and-effect relationships. They write cause-and-effect statements about what they've observed about the relationship between physical properties, phase changes, and heat energy.

OBJECTIVES

- Name and describe physical properties of solids, liquids, and gases
- Explain how temperature changes during melting, freezing, and boiling
- Explain that energy (heat) changes the physical properties of a substance
- Write sentences demonstrating cause-and-effect relationships

STANDARDS ALIGNMENT

National Science Education Standards
SCIENCE AS INQUIRY

- K–4, 5–8 Abilities Necessary to Do Scientific Inquiry

PHYSICAL SCIENCE

- K–4 Properties of Objects and Materials
- 5–8 Properties and Changes of Properties in Matter

SCIENCE AND TECHNOLOGY

- K–4, 5–8 Abilities of Technological Design

Common Core State Standards for English Language Arts

INFORMATIONAL TEXT

- Grades 3–5 Key Ideas and Details
- Grade 3 Integration of Knowledge and Ideas

WRITING

- Grades 3–5 Text Types and Purposes

For a detailed standards alignment, see Appendix 3 (p. 282).

TIME FRAME

- Nine 45-minute class periods

SCIENTIFIC BACKGROUND INFORMATION

Solids, liquids, and gases are three of the states of matter. (The others—plasmas and the Bose-Einstein condensates—are more complex subjects than would be appropriate for the grade-level expectations of elementary students.) Each state (or phase) of matter has a set of general properties. Solids have a definite shape and volume. They do not flow easily and are not easily compressible. Liquids do not have a definite shape, but instead take the shape of their container. They do have a definite volume. They do flow easily, but are not easily compressible. Gases assume the shape and volume of their container. They flow easily and are easily compressed.

The differences in properties between phases can be explained by considering the molecular arrangement in each. In a solid, these molecules are packed tightly together. Since there is little space between them, the solid is not easily compressed and does not flow. The rigid arrangement of the molecules also explains why a solid retains its shape and volume, no matter what container it is placed in. In a liquid, the molecules are able to move past each other, which makes the liquid able to flow and to take the shape of its container. There is still little space between the molecules, making it difficult to compress. And finally, in a gas, the molecules move freely. This means that a gas takes the shape of its container, has no definite volume, is easily compressible, and flows easily. It is beyond grade-level expectations for elementary students to understand the molecular motion and structure of solids, liquids, and gases. Instead, students should focus on the observable properties of each state of matter and begin to consider how energy plays a role in changing from one state to another.

When solids turn to liquids (or liquids to gas), their molecules gain more energy and thus move more freely. Typically, this happens when materials are heated. Ice cubes left out of the freezer will melt. We melt butter or chocolate in a saucepan on the stove, or in the microwave. Coconut oil is a solid at temperatures below 76°F and a liquid above this point. All of these things happen because the molecules of the material in question gain energy in the form of heat. The melting point and the freezing point of a substance are actually the same temperature. Whether the substance melts or freezes depends on whether energy is increasing or decreasing.

We often demonstrate a liquid turning to a gas by boiling water for students. However, it is important to realize that boiling and evaporation are not the same. Evaporation takes place at the surface of the liquid, whereas boiling occurs throughout the liquid. Evaporation can occur at any temperature, whereas boiling occurs at a temperature specific to that liquid (known as the boiling point). The steam rising from a pot of boiling water is often referred to (incorrectly) as water vapor. Water vapor cannot be seen. Bubbles in boiling water consist of water vapor, but that vapor condenses into tiny droplets that we see as steam rising from the container.

MISCONCEPTIONS

Phase changes are a difficult concept, and students may have scientifically inaccurate ideas, or misconceptions about melting, freezing, evaporation, and condensation. Table 8.1 lists some documented misconceptions along with the scientifically accurate concepts (Bar 1985; Henriques 2000).

TABLE 8.1. COMMON MISCONCEPTIONS ABOUT MATTER AND PHASE CHANGES

Common Misconception	Scientifically Accurate Concept
The bubbles in boiling water are air, oxygen, hydrogen, or heat.	The bubbles in boiling water are water vapor.
What we observe coming from a pot of boiling water is smoke or hot air.	Steam can be observed rising from a pot of boiling water. Water vapor condenses to form tiny droplets of water, which we see as steam.
Water in an open container disappears, changes into air, or is absorbed by the container.	Water in an open container evaporates. It changes from a liquid to a gas.
Condensation on the outside of a container (like a cold drink) is water that has seeped through the container.	When the water vapor in the air comes into contact with a cool surface, it condenses into water droplets. These water droplets form the condensation we see on the outside of containers.
When water evaporates, it ceases to exist.	When water evaporates, it changes from a liquid to a gas.

As students observe the phase changes during the explore phase of the unit, listen carefully to their explanations in discussion and in writing. Do they understand that air (a gas) is an actual state of matter? In particular, accepting air (gas) as a permanent substance as opposed to nothingness is a challenging concept for many students (Sere 1985).

TEXT SET

Ice Cream: The Full Scoop by Gail Gibbons (New York: Holiday House, 2006); reference and explanation, Flesch-Kincaid reading level 4.4.

Who doesn't love to eat ice cream? Learn about the history of this sweet frozen treat and how it is made. Includes a wonderful cross section of an ice cream maker.

Ice to Steam: Changing States of Matter by Penny Johnson (Vero Beach, FL: Rourke, 2008); reference, Flesch-Kincaid reading level 4.4.

Full-color illustrations and clear diagrams accompany this accessible text about states and changes of matter. Eight chapters cover various concepts, including changes of water, volcanoes and lava, cooking, and solids that behave as liquids.

A Look at Glaciers by Patrick Allen (New York: Rosen, 2009); reference, Flesch-Kincaid reading level 5.4.

What is a glacier, and how does it change the landscape? Learn all about these massive rivers of ice with short chapters, colorful pictures, and clear diagrams.

Many Kinds of Matter: A Look at Solids, Liquids, and Gases by Jennifer Boothroyd (Minneapolis, MN: Lerner Publications, 2011); reference, Flesch-Kincaid reading level 3.2.

Learn about matter, the properties of solids, liquids, and gases, and how matter changes in this colorful book.

MATERIALS

- Resealable plastic sandwich bags (one per pair of students)
- Whole or 2% milk (½ c. per pair of students) or juice
- Vanilla extract
- Sugar
- Salt
- Paper or plastic cups
- Plastic spoons
- A variety of solids: ice, stick of butter, chocolate, coconut oil, marbles, chocolate chips
- A variety of liquids: water, melted butter, melted chocolate, melted coconut oil, molasses or honey, rubbing alcohol
- Helium balloons and regular balloons
- Word wall or chart paper to create a temporary word wall
- Markers
- Crayons and pens
- Glass containers, such as beakers
- Containers of a variety of sizes and shapes, including bowls
- Balance or scale
- Jelly jar or mason jar
- Bubble wrap
- Rubber band or string long enough to tie around the jelly or mason jar
- Student science journals
- Hot plate
- Wax paper
- Eyedropper
- A variety of books about the polar regions
- Shoe boxes of identical sizes
- Cardboard, Styrofoam, cotton balls, newspaper, and other items that could be used as insulators
- Packing tape
- Large ice cubes or blocks of ice (such as ice cubes made by freezing water in milk cartons)

- Temperature probe (optional)
- Copies of supporting documents

SUPPORTING DOCUMENTS

- Three-circle Venn diagram
- Venn diagram answer key
- "Physical Properties Graphic Organizer"

SAFETY CONSIDERATIONS

Students with milk allergies or diabetes should not eat the ice cream. Consider having a safe alternative on hand for them. Juice can be used in place of the milk to create a "slushie" instead of ice cream. If juice is used, omit the vanilla and sugar described in step 3 of the engage phase.

Minds-On Matter Inquiry Unit

ENGAGE

In the engage phase, students will experience a tasty example of a phase change as they make ice cream. The experience will prompt them to think about the properties of materials and how they can change.

1. Read the first 11 pages of *Ice Cream*. Tell students that they don't need an ice cream maker—they can make their own ice cream right here in the classroom!

2. Explain the process to students (described below), holding up the ingredients one at a time. Ask them to describe the ingredients (milk, sugar, vanilla extract, ice). Record student observations on chart paper in a column labeled "Before."

3. Now it's time to make the ice cream! Students will work in pairs for this activity. In a small (sandwich-size) resealable plastic bag, combine ½ c. of milk (whole or 2% is best), ½ tsp. of vanilla extract, and 1½ tbsp. of sugar. Seal the bag well. Next, fill a gallon-size resealable plastic bag with ice. Add 6 tbsp. of salt to the ice. Place the small bag inside the larger bag, and seal. Shake the bag vigorously for approximately five minutes. The bags get cold, so students may want to wear gloves while shaking the bags. *Note:* This activity can be messy, so you may want to conduct this outside if possible! Parent volunteers can also be helpful.

4. After five minutes of shaking, have students open the bags. What has happened to the ice? How would students describe it now? Remove the small bag, carefully wipe off the top, and open it. What has happened to the milk? How would students describe it now? Record student observations on chart paper in a column labeled "After."

5. Scoop the ice cream into individual paper or plastic cups, pass out spoons, and let students enjoy! As they are eating their ice cream, ask students what they have experienced, observed, and tasted. Explain that what they observed was a change from a liquid to a solid, or freezing. *Freezing* is an example of a phase change. Ask students if they know any other examples of phase change. They may name *melting, evaporating,* or *condensing*. Add these phase change words to your word wall. Explain to students that, as they observed with the ice cream, some of the properties, or characteristics of materials change during a phase change. Introduce the unit question: *How do the physical properties of a material change during phase changes?*

Assess this phase: At this stage of the inquiry, formative assessment should be the focus. Listen carefully to the words and phrases students use to describe the ingredients and ice cream. This will provide insight into students' current understanding of physical properties.

EXPLORE

In the explore phase, students will explore properties of solids, liquids, and gases. They will also observe phase changes as ice melts and water boils.

1. Discuss the idea of *physical properties* (observable characteristics) with students, and add the term to the word wall. With student assistance, generate a list of properties that they can observe in the classroom. This list should include temperature, shape, color, and mass. Students may suggest other physical properties, but these should be listed at a minimum. Also discuss how to measure, observe, and record the named properties.

2. Provide time for students to investigate the properties of a variety of solid objects (see Materials list for suggestions) in small groups or pairs, as materials permit. Students should have a chance to measure the mass, to observe the object's shape and if it changes when moved from one container to another, and to draw and describe its color. Students should record their measurements and observations in a data table like the one shown in Table 8.2. Students can construct these tables in their science journals, or you can provide a blank copy of the "Physical Properties Graphic Organizer" for students to complete.

TABLE 8.2. SAMPLE DATA TABLE FOR SCIENCE NOTEBOOKS

Object	Mass	Shape	Does Its Shape Change?	Color

3. Once students have had an opportunity to explore a variety of solids, read pages 6–9 of *Many Kinds of Matter* aloud. (Alternatively, if you can find enough copies of the text, consider having students read those pages in their small groups.) Ask students how their findings compare with the text. These guiding questions may be helpful:

    *   Did the text confirm what you observed about the solids? What parts confirmed your observations?

    *   What new information did you learn about solids?

    *   Would you like to explore anything else after listening to the text?

Students should record any new information that they learn about solids. Allow students to conduct further exploration with the solids if the text has raised any issues or questions.

Next, students will observe how the temperature changes as ice cubes melt and turn to liquid water. Conduct this portion as a whole-class activity.

4. Fill a beaker or other clear glass container with ice cubes. Place a thermometer or temperature probe into each container, being sure that the thermometer or probe is not resting on the bottom of the container. Students should record this and all temperature data in their science journals. Continue to record the temperature every three minutes. Although the ice will melt, it will take quite some time.

5. After 15 or 20 minutes, ask students if they can think of a faster way to make the ice melt. They may suggest putting the ice in the sun or on a heat source. Follow their suggestion and set up a second beaker (or glass container) full of ice cubes. Record the temperature of this second beaker every three minutes while in the sun or on a heat source, and continue to measure the temperature of the unheated ice. During this time, ask students to periodically describe what they are observing. How is the ice changing? Which properties are changing? Discuss that the ice has melted and is now liquid water. Which container's ice melted fastest? Why?

    Note: If you have access to a temperature probe, we highly recommend it for this activity. Probes will often record the data and generate a graph, helping students make the connection between their observations, the data, and the graph that represents it. Students can still practice creating a graph, but having one to view provides a nice scaffold while they master this skill.

6. Create a line graph with students of the temperature over time, using two colored lines to represent the two containers of ice. Alternatively, students can graph the data independently in their science journals.

7. Place a container with chocolate chips on a hot plate. Place a thermometer or temperature probe into the container, taking care that the thermometer or probe is not resting on the bottom of the container. Students should record the temperature at three-minute intervals until the chocolate has melted. Add the data to the line graph using a third color to represent the chocolate. Engage students in a discussion about the data. How did the chocolate compare with the ice in terms of melting? What does this tell students about the melting process?

8. Remove the chocolate from the heat source once it is melted, and let it sit undisturbed. Once it has hardened, show it to students and discuss how this is the opposite of melting. Even though we don't normally refer to this as freezing, it is still an example of freezing. This provides an opportunity to discuss the differences between scientific language and common language. It also helps students consider freezing in terms of materials other than water. Encourage students to make connections to the ice cream that students made during the engage phase, and how the ice cream could be refrozen by placing it in the freezer.

9. At this time, students will return to small groups to investigate a variety of liquids (see Materials list for suggestions) and their properties. It is important to provide liquid versions of some of the solids to allow students to make direct comparisons. Students should measure the mass of the liquids and describe the colors they observe. They should also have an opportunity to pour liquids from one container to another to learn that liquids take the shape of their containers. Students will record their observations in a data table similar to the one used in the solids investigation.

10. Read pages 10–13 of *Many Kinds of Matter* aloud. (Alternatively, if you have enough copies of the text, consider having students read those pages in their small groups.) After reading these pages, ask students how their findings compare with the text. These guiding questions may be helpful:

    •   Did the text confirm what you observed about the liquids? What parts confirmed your observations?

    •   What new information did you learn about liquids?

    •   Would you like to explore anything else after listening to the text?

    Students should record any new information that they learn about liquids. Allow students to conduct further exploration with the liquids if the text has raised any issues or questions.

    Next, students will observe how the temperature changes as water heats and boils. They will also begin observing water before and after evaporation. Conduct these portions as a whole-class activity.

11. Use an eyedropper to create a puddle of water on a piece of wax paper. Approximately 25 drops is a sufficient size. Create a second puddle with rubbing alcohol. Draw outlines around each puddle with a pen or a crayon. Leave this undisturbed in the classroom overnight. Also fill two containers with water and rubbing alcohol and record the initial level in each. Place a thermometer into each container and record the temperature.

    Note: Evaporation is extremely dependent on environmental conditions, such as the humidity. Test this portion of the activity before conducting it in class, and adjust the water level and puddle size to ensure that students will be able to observe a noticeable difference after a 24-hour period.

12. Place the container of water (the same one used in the previous observation of ice melting) on a heat source, such as a hot plate. Turn on the heat source and record the initial temperature. Continue to record the temperature every three minutes until water is boiling rapidly. During this time, ask students to periodically describe what they are observing. How is the water changing? Which properties are changing? Discuss that as the water boils, it is turning into a gas. Although the water isn't evaporating, it is still changing state. However, the steam that can be observed rising from the container is not water vapor—it is still liquid water. Water vapor cannot be seen.

13. Create a line graph of the temperature over time with students, or have students graph the data independently in their science journals.

14. The next day, observe the puddle that was left undisturbed overnight. Draw a second outline of the puddle on the wax paper. How has the size of the puddle changed? What has happened to the water and rubbing alcohol? Observe the water level in the container and record the temperature again. Has the water level changed? What about the temperature? Discuss that the water and rubbing alcohol evaporated overnight, leading students to conclude that liquids other than water evaporate.

15. Students will return to small groups to investigate a variety of gases and their properties. They will compare the mass of a filled and an unfilled balloon to investigate whether or not air has mass. They will also blow up balloons and let them go to observe that a gas does not maintain a fixed shape or volume. Including helium balloons will help students generalize beyond air. *Note:* If you do use helium balloons, we recommend purchasing them the morning that you will use them. They deflate quickly!

16. Students can also pop bubble wrap and test the idea that a gas takes up space by placing a jelly jar or mason jar in a plastic bag and tying the bag tightly. Can they push the bag into the container? During these activities, they will record observations about the properties of a gas in their science journals.

17. Read pages 14–17 of *Many Kinds of Matter* aloud. (Alternatively, if you have enough copies of the text, consider having students read those pages in their small groups.) After reading these pages, ask students how their findings compare with the text. These guiding questions may be helpful:

    • Did the text confirm what you observed about gases? What parts confirmed your observations?

    • What new information did you learn about gases?

    • Would you like to explore anything else after listening to the text?

    Students should record any new information that they learn about gases. Allow students to conduct further exploration with air if the text has raised any issues or questions.

18. Read pages 4–7 of *Ice to Steam* aloud. After reading, ask students to relate the text to what they observed and graphed. How do their findings and observations compare with what was said in the text? If needed, repeat the process of melting ice and then boiling the water while monitoring the temperature.

Assess this phase: Formative assessment should be used throughout this phase. Observe students as they investigate the properties of the solids, liquids, and gases. Use guiding questions like "What do you observe about the (object)?" to help students focus on the relevant physical properties. Guiding questions will also help ensure that students connect what they observe firsthand to what they read in the text. If students struggle to identify physical properties of any of the three states of matter, provide additional time for exploration with structured questions. Additionally, student responses while observing the melting and boiling processes will provide insight into students' understanding of how materials change from one state to another and how physical properties change with phase changes. If students seem to be struggling, increase the number of guiding questions asked. Also consider repeating the phase change demonstrations after reading the passage from *Ice to Steam*.

EXPLAIN

In the explain phase, students will use what they've learned from hands-on experiences and text to generalize about the properties of solids, liquids, and gases. They will also explain how those properties change with phase changes.

1. Have a brief conversation about how the properties of solids, liquids, and gases compare. You may want to list properties of each state of matter on the board or on chart paper.

2. Students will complete a three-circle Venn diagram, comparing and contrasting the properties of water as a solid, liquid, and gas.

3. Introduce cause-and-effect text. Reread pages 6–7 of *Ice to Steam* aloud, highlighting sentences and phrases that signal cause-and-effect relationships. If possible, project the text or copy sentences onto the board. Ask students what cause-and-effect relationships they observed during these investigations. Students should explain that adding heat to the ice caused the ice to melt and that adding heat to the water caused the water to boil. Discuss the melting and evaporating that happened without a heat source.

4. Help students dig deeper into this topic by asking guiding questions like:

    *   How does adding heat (energy) to a solid change its properties?
    *   What about adding heat (energy) to a liquid?
    *   Did you observe any changes in properties that didn't need energy?
    *   Is energy necessary for phase changes?
    *   Do phase changes mean that physical properties change?

5. Students will share what they've learned by writing cause-and-effect statements about the materials they observed, their properties, and the changes observed during phase changes. Students should have access to their science journals, the materials, and the texts while they write these statements. You may also wish to provide sentence stems to support students' writing, particularly if they are less familiar with cause-and-effect text structure:

- If _____ then _____
  If I add heat to a solid, then it melts.

- _____ so _____
  The ice melted so it didn't stay the same shape.

- Because _____, _____.
  Because I left the water out overnight, it evaporated.

- _____ and _____
  The ice melted and it didn't stay the same shape.

Assess this phase: The three-circle Venn diagram and cause-and-effect sentences serve as summative assessment for this inquiry. The Venn diagram provides evidence of students' ability to name and compare properties of solids, liquids, and gases, while the sentences provide evidence of how properties change with phase changes. We've provided an answer key for the Venn diagram, but other responses may be acceptable. The "Science and Literacy Rubric" in Appendix 2 (p. 276) can be used to assess student performance, and the "Achievement Grading Standards" (also found in Appendix 2 [p. 277]) can translate this into a numerical grade. If students fall below 75%, return to the explore phase for additional work before returning to the writing prompt.

EXPAND

In the expand phase, students will apply what they've learned about physical properties and phase changes to design a container to transport ice without it melting.

1. Explain to students that they are going to study a part of the world with a lot of ice—the polar regions. Provide a variety of books about the polar regions and allow students to browse and skim them. Talk about what they found in the text and pictures. What examples of water as a solid did they find? Students should name ice, snow, icebergs, and glaciers.

2. Read pages 24–25 of *A Look at Glaciers*, which discusses how scientists study glaciers by drilling for ice cores. The article "Unlocking the Climate History Captured in Ice" (*http://bit.ly/sz3FK*

QR Code 1

[QR Code 1]) includes pictures and video of ice cores being taken and transported in the field. Explain that scientists have to be careful to transport the ice cores across the world without letting them melt. Introduce a design challenge task to students: Design a container that will prevent an ice cube from melting.

3. Before students design their containers, review what students have learned about melting. Why would an ice cube melt? (It gains energy, usually in the form of heat.) How can you keep an object cold? Students should be able to name coolers and insulated lunch boxes as examples of products that keep objects cold. Introduce and define the words *insulation* and *insulator* and add these to the classroom word wall.

4. Have students conduct a brief experiment to test the insulating qualities of various materials, such as cardboard, Styrofoam, cotton balls, and newspaper. To do this, fill a beaker with ice and place it inside a shoe box filled with one of the materials. Seal the box and wait 20 minutes. After 20 minutes unseal the box and measure the remaining ice. Repeat for each material, or conduct the tests simultaneously if you have enough containers. Be sure to use the same-size beaker in each trial, put the same quantity of ice in each beaker, and use the same-size shoe box for each trial.

5. Teams of students will design the containers, using materials provided in the classroom (or brought from home, at your discretion). Teams should prepare a presentation that describes how they plan to build the prototype. The presentation should include diagrams that illustrate the finished product. Refer students to the cross section of the ice cream maker on page 9 of *Ice Cream* as an exemplar diagram.

6. Teams will briefly present their plans and diagrams to the class. Other students should ask questions and provide feedback about strengths and suggestions for improvement. The teams will now have a chance to modify their plans, if needed.

7. At this point, students will build their containers. Once containers are built, provide teams with an ice cube, which will represent an ice core. Students should measure the volume and mass of the ice cube before placing it inside their container. The next day students should open the containers, measure how much liquid water is present in the container, and share their findings with the class. Teams should prepare a brief report including their design and diagram, their results, and suggestions for improvement.

Assess this phase: At this point, assessment focuses on designs for containers and student presentations of their work. This is a time to assess science process skills: controlling variables, formulating a hypothesis, experimentation, and interpreting data. Students are not only expanding their understanding of physical properties and phase changes but also developing their expertise with these process skills. Assessment should be focused not on a correct answer but on the process skills themselves. We recommend using the "Science Process Skills Rubric" in Appendix 2 (p. 277) to assess student abilities in this area.

REFERENCES

Bar, V. 1985. Children's views about the water cycle. *Science Education* 73 (4): 481–500.

Henriques, L. 2000. Children's misconceptions about weather: A review of the literature. Paper presented at the National Association of Research in Science Teaching annual meeting, New Orleans. *www.csulb. edu/~lhenriqu/NARST2000.htm*

Sere, M. 1985. The gaseous state. In *Children's ideas in science*, ed. R. Driver, 105–123. Buckingham, England: Open University Press.

Name_____ Date_____

VENN DIAGRAM

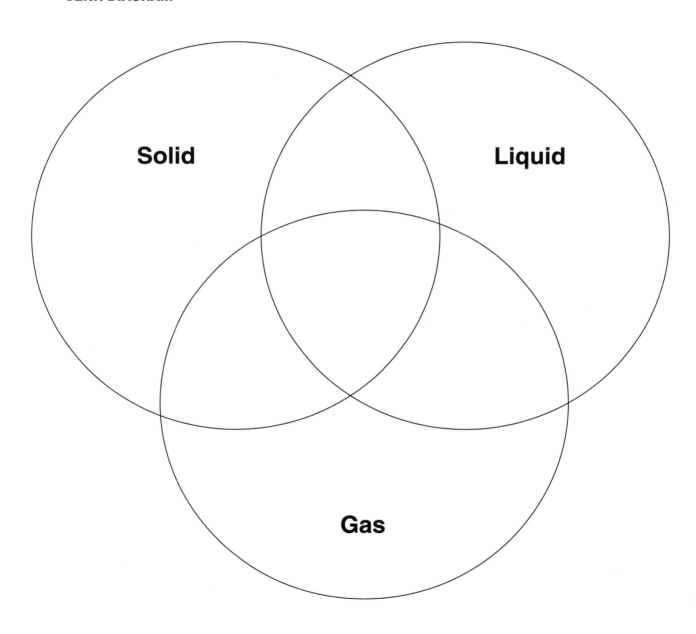

Name_____ Date_____

COMPLETED VENN DIAGRAM

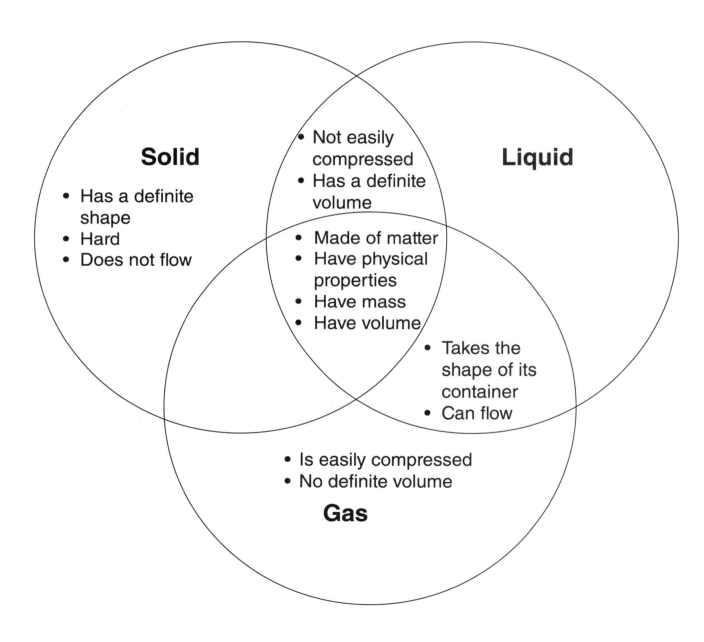

Name_____ Date_____

PHYSICAL PROPERTIES GRAPHIC ORGANIZER

Object	Mass	Shape	Does Its Shape Change?	Color
What I Learned From the Text				

Chapter 9

Classroom Curling: Exploring Forces and Motion

OVERVIEW

Through this inquiry, students work to answer the question, *What happens to the motion of a slider when different forces are applied?* Teams of students play a game of "classroom curling"—a loose adaptation of the sport of curling. During the game, students try to change the direction of an object, making the game a natural lead-in to investigations into forces and motion, the focus of this unit. Students connect what they have learned to the activities described in the book *Newton and Me*. To close the inquiry, students apply what they have learned to improve their classroom curling skills.

This inquiry unit gives students the opportunity to use text features such as a table of contents and index to locate information. Students also synthesize information from hands-on experiences and nonfiction text to define content-specific vocabulary, make text-to-self connections, and write and illustrate (with diagrams) an explanatory text.

OBJECTIVES

- Explain that forces are required to start, stop, or change the direction of an object's motion
- Investigate and define friction
- Investigate and define gravity
- Use text features to locate information in nonfiction texts
- Write and illustrate an explanatory report
- Collect and analyze data to make evidence-based claims
- Make text-to-self connections with nonfiction text

STANDARDS ALIGNMENT

National Science Education Standards
SCIENCE AS INQUIRY

- K–4, 5–8 Abilities Necessary to Do Scientific Inquiry
- K–4 Understanding About Scientific Inquiry

PHYSICAL SCIENCE

- K–4 Position and Motion of Objects
- 5–8 Motions and Forces

Common Core State Standards for English Language Arts
INFORMATIONAL TEXT

- Grades 3–5 Craft and Structure

WRITING

- Grades 3–5 Text Types and Purposes

For a detailed standards alignment, see Appendix 3 (p. 282).

TIME FRAME

- Seven to nine 45-minute class periods

SCIENTIFIC BACKGROUND INFORMATION

A *force* is generally defined as a push or pull. Forces are required to set an object in motion, change its direction, or stop its motion. Newton's first law of motion tells us that an object at rest stays at rest and an object in motion stays in motion in a straight line at a constant speed unless acted upon by an external force. The only way to move something that is still is to apply a force, to push or pull it. Likewise, the only way to stop an object that is in motion is to apply a force, to push or pull in the opposite direction of the motion. At this grade level it is not necessary for students to state Newton's first law of motion. What you want students to understand is that forces are required to start, stop, and change the direction of an object's motion.

Two forces that are always present on Earth are gravity and friction. Gravity pulls objects toward Earth. When an object is tossed into the air, gravity's pull slows the object even as it is rising in the air. At this point, gravity is pulling in the opposite direction of the object's motion. When the object begins to fall back to Earth, the force of gravity is acting in the same direction of the motion. This results in an increased velocity in a downward direction. We see the force of gravity in action when an object slides down a ramp. Gravity pulls on the object as it travels down the ramp. At this grade level, what you want your students to understand is that gravity is a force that pulls objects toward Earth.

Friction is a force that acts in the opposite direction to an object's motion. Friction slows objects, but it also works to hold objects in place. To set an object in motion, the force has to be great enough to overcome the weight of the object and the force of friction that holds the object in place. Frictional forces act in the opposite direction of an object's motion, slowing it down. Friction exists everywhere. It occurs when two objects are in contact with each other. The smoother the object, the lower the frictional force. The big idea here for students is that friction is present everywhere, holds objects in place, and slows the motion of objects.

When a force sets a stationary object in motion, the object accelerates in the direction of the force. The greater the force, the greater the acceleration. This is one part of Newton's second law of motion. The mass of an object in relation to force is the other part. Newton's second law tells us that acceleration is directly proportional to force and inversely proportional to mass. This means that:

- If the same force is applied to two objects with different masses, the objects will accelerate at different rates. The object with the larger mass will have a lower rate of acceleration. The one with the smaller mass will have a greater rate of acceleration.

- If different forces are applied to two objects with the same mass, the objects will have different rates of acceleration. A larger force will result in greater acceleration than a smaller force.

It is not necessary for students to state Newton's second law. What you want them to come away with is an understanding that larger forces cause objects to move faster than do smaller forces and that heavy objects require more force to move than lighter objects.

MISCONCEPTIONS

Misconceptions about forces and motion are often based on everyday experiences. Driver et al. (1994) compiled a list of misconceptions about forces and motion that reveal children's thinking about this concept. We have restated the ones relevant to this unit in Table 9.1 (p. 114).

Along with these misconceptions is the confusion that comes from the daily use of the word *force*. Children are likely to associate the word *force* with coercion, physical activity, or muscular strength. They may view a force as evidence of an inanimate object "trying" to accomplish something (Driver et al. 1994). It's an opportune time to discuss the difference between the daily and scientific use of the term after reading *Exploring Forces and Movement* in the engage phase. You can further establish a scientific way of thinking about forces and their role in motion by focusing students' attention on the role of forces in the motion of the sliders throughout the unit.

TABLE 9.1. MISCONCEPTIONS ABOUT FORCES AND MOTION

Common Misconception	Scientifically Accurate Concept
If there is motion, a force is acting.	Although a force is needed to set an object in motion, no force is needed to keep it in motion.
No force is acting on a stationary object. Forces only exist when objects are in motion.	Forces are always present. Balanced forces act on stationary objects. These forces act in opposite directions and are of equal strength.
A moving object stops when its force is used up. A moving object has a force within it.	When a force is no longer being applied, its effect stops. Forces are not tangible materials that can be used up. Forces are external to moving objects, they slow or stop the motion of an object.
Motion is proportional to the force acting on the object.	Motion is a change in the position of an object. A force is required to start, stop, or change the direction of an object in motion. Acceleration, the change in an object's speed or direction, is proportional to force.
When an object is moving, a force is pushing it in the direction of its motion.	A force is required to start, stop, or change the direction of an object's motion. But once in motion, an object will continue in motion in a straight line at a constant speed. No force is required to sustain the motion.
Forces make things go and are not associated with making things stop.	Forces are required to change an object's motion. This includes starting, stopping, and changing direction.

TEXT SET

Exploring Forces and Movement by Carol Ballard (New York: PowerKids Press, 2008); explanation, Flesch-Kincaid reading level 4.6.

Readers learn about forces, gravity, friction, and more in this engaging book. The "Try This!" activities give students the opportunity to try their hand at activities such as finding a balance point and making a spinning force. Photos capture children in the midst of an activity.

Forces: The Ups and Downs by Wendy Sadler (Chicago: Raintree, 2006); explanation, Flesch-Kincaid reading level 4.5.

Students will find information about forces including friction, gravity, drag, magnetism, and forces in the body. The text is clear and well written, and the brief annotations that accompany the many photographs play an invaluable part in explaining the concepts. Even the text font used in the book is pleasing.

Ice Hockey and Curling by Robin Johnson (New York: Crabtree, 2010); reference, Flesch-Kincaid reading level 6.6.

This is what we call a "snack" book. The information is presented in bite-size portions, which makes it entertaining to read, especially for reluctant readers. Each set of pages presents a different topic. Put them all together, and you get a solid overview of ice hockey and curling—as played by both men and women.

Motion by Darlene Stille (Chanhassen, MN: Child's World, 2005); explanation, Flesch-Kincaid reading level 4.5.

This book provides information about how things start, stop, and change their motion; photos that support the text appear on almost every page.

Newton and Me by Lynne Mayer (Mt. Pleasant, SC: Sylvan Dell, 2010); narrative expository, Flesch-Kincaid reading level 2.7.

Follow the adventures of a boy and his dog, Newton, as they encounter forces and motion during their daily activities—and they do it in rhyme! Fun drawings by Sherry Rogers illustrate the text.

What Is a Force? by Jacqui Bailey (Mankato, MN: Smart Apple Media, 2005); explanation and reference, Flesch-Kincaid reading level 4.4.

You will find the answer to the title question through clearly written text and a series of simple activities that readers can do on their own. Numerous lively color photographs show children performing the activities.

MATERIALS

- Masking tape
- Chalk, wax pencil, or other nonpermanent writing instrument
- Curling sliders, four per team (You can use furniture sliders, which are available at home improvement stores and discount department stores.)
- Bumpers (You need objects that students can place in the classroom curling playing area to redirect the sliders; board erasers work well.)
- Ramp materials (must be very smooth and about 75–100 cm long)
- Supports for the ramp (textbooks work well)
- Printer paper
- Wax paper
- Sandpaper (120 grit)
- Stopwatch
- String
- Snack-size plastic bag
- Weights (If calibrated weights are not available, any items of equal weight will work, such as washers or marbles.)
- Graph paper
- Copies of supporting documents

SUPPORTING DOCUMENTS

- "Changing Force I" data sheet
- "Changing Force II" data sheet
- "Defining Friction" graphic organizer
- "Defining Gravity" graphic organizer
- "Newton and Me Connections" graphic organizer
- "Putting the Pieces Together Report Planner"

SAFETY CONSIDERATIONS

Students should be careful not to use too much force when pushing the sliders. Have team members and observers stand clear of the path of the sliders.

Classroom Curling Inquiry Unit

CLASSROOM CURLING GAME DESCRIPTION

Objective

Land your slider as close as possible to the button, or the center of the "house."

Playing Area

1. Clear an area of the floor that is approximately 3 m long and 1 m wide for the playing area. Figure 9.1 illustrates the playing area. This is a suggested playing area. The dimensions are not as important as the general layout. Change the dimensions to fit the needs of your classroom. In fact, the playing area could be on a tabletop if necessary.

FIGURE 9.1. CLASSROOM CURLING PLAYING AREA

2. Place a piece of masking tape at one end of the playing area for the hogline.

3. Place another piece of masking tape at the opposite end for the backline.

4. Using chalk, a wax pencil, or other nonpermanent means, draw two concentric circles near the backline. The largest circle, the house, should be about 60 cm in diameter. The smallest circle, the button, should be about 15 cm in diameter.

5. Place a third piece of tape in front of the concentric circles for the teeline.

6. Allow extra space all the way around the playing area for the players to maneuver.

Team

Each team has four members, and each member has a specific role.

- Lead: Throws (slides the curling slider) first; bumps (strategically places a bumper between the hogline and teeline and in the path of the curling slider in an effort to direct its motion) for all other players (Second, Vice, and Skip).
- Second: Throws second; bumps for all other players (Lead, Vice, and Skip).
- Vice: Throws third; bumps Lead's and Second's shots
- Skip, team captain: Throws last; tells other team members when to bump.

Equipment

- Sliders and bumpers

Rules

1. Two teams compete against each other in an effort to slide their sliders as close as possible to the button (center circle).

2. Each team member throws on his or her turn, releasing the slider anywhere in front of the hogline. Before the slider is in motion, two team members may bump. Bumping may not be needed if the most strategic path for the slider is a straight line. The bumpers are the only things that can touch the slider when it is in play.

3. Teams take turns throwing.

4. The Skip stands behind the backline and advises the other players on the path and the placement of the bumpers.

5. Sliders completely beyond the backline are out of play.

6. Teams can knock one another's sliders out of position with their slider to score more points.

Play

1. Flip a coin, draw straws, or use some other method to decide which team goes first.

2. Order of play:
   - Round 1: The Second and Vice for the team going first bump. The Lead for the team throws. The bumpers are removed from the playing area. The other team then takes a turn.
   - Round 2: The Lead and Vice bump. The Second throws. The bumpers are removed and other team then takes a turn.
   - Round 3: The Lead and the Second bump. The Vice throws, the bumpers are removed, and the other team then takes a turn.

- Round 4: The Lead and the Second bump. The Skip throws, the bumpers are removed, and other team takes the final turn.

Scoring

All sliders that are completely or partially in the house score 3 points. Sliders with all or part of the slider in the button are given 5 points. Sliders between the teeline and the backline get 1 point. Sliders beyond the backline are out of bounds and do not get any points. The team with the most points wins.

ENGAGE

In the engage phase, students play a game that requires thoughtful consideration of the forces and motion of a sliding object.

Advance preparation: Set up the classroom curling playing area (as described earlier in this unit) before beginning the inquiry.

1. Show one or more of the following video clips to introduce the sport of curling. QR codes for the videos have been provided for your convenience. Use a scanning app on your smartphone, on your tablet, or with the webcam on your computer to scan and quickly access the videos.

   - "Jennifer Jones Best Curling Shot!": *www.youtube.com/watch?v=CM5mFH3_Qhs* (QR Code 1)
   - "Curling Is INTENSE!!!!!": *www.youtube.com/watch?v=3C-__7sAePs* (QR Code 2)
   - "Curling Rocks": *www.youtube.com/watch?v=idSdnubrlds* (QR Code 3)

QR CODE 1   QR CODE 2   QR CODE 3

2. Read aloud pages 20–21 of *Ice Hockey and Curling* to provide additional information about the sport of curling.

3. Share with the students that they will be taking part in a classroom version of curling. Explain that the classroom version is like real curling in that the teams will be made up of four players, teams will slide an object toward a target, team members will be assigned similar roles, and the scoring and rules will be similar. The main differences are that in classroom curling a slider will be used in place of the stone, players will be bumping instead of sweeping, four sliders will be used instead of eight, and the surface will be the classroom floor instead of an ice rink.

4. Demonstrate how classroom curling is played. Let students practice sliding the slider.

5. Divide the class into teams of four. Have each team play at least one game of classroom curling. Students who are not actively playing should pay attention to the techniques used by the competing teams. As the students play, ask guiding questions such as:

   • What were some of the best techniques for hitting the button?
   • What did you notice when you were sliding the slider?
   • What happened when the slider came into contact with a bumper?

6. After play comes to an end, discuss with the students which techniques worked the best.

7. Summarize the discussion and introduce the question, What happens to the motion of a slider when different forces are applied?

8. Read aloud pages 4–6 of *Exploring Forces and Movement* to introduce the concept of a force.

Assess this phase: Formative assessment should be used during this phase. Observe students playing the game, and listen carefully to their responses to your guiding questions. These responses will give you an idea of students' preconceptions about forces and motion.

EXPLORE

In this phase students will work in small groups to investigate how forces speed up, slow down, and change the direction of objects in motion.

Changing Direction

In this part of the explore phase, students demonstrate that a slider travels in a straight path unless it collides with a bumper.

1. Students will first confirm that the slider travels in a straight path if no external force acts on it. Put the slider on a smooth surface (e.g., the floor or a tabletop). Give it a push with a moderate amount of force. The slider should travel in a fairly straight path.

2. Model for the class how to diagram the path the slider follows by drawing it on the board. Students can replicate the diagram in their science notebooks. Students will create additional diagrams later in the unit.

3. Discuss with the students the types of information the diagram illustrates.

4. Use these guiding questions to summarize the activity and transition to the next steps:

   • What do you notice about the path of the slider? (Is it straight, curved, zigzagged?)

- How might you change the direction of the slider?
- What would happen if you pushed the slider harder? More gently?

5. Demonstrate that when a slider collides with a bumper, it changes direction. Students will have observed this when they played classroom curling. Demonstrating it here provides an opportunity to model how to diagram the slider's path. To demonstrate:

    a. Set a bumper at an angle in the path of the slider as shown in Figure 9.2.

    b. Diagram the setup on the board.

    c. Ask students to predict what will happen when the slider collides with the bumper.

    d. Launch the slider so that it reflects off the bumper when they collide.

    e. Diagram the path of the slider.

FIGURE 9.2. PATH OF THE SLIDER

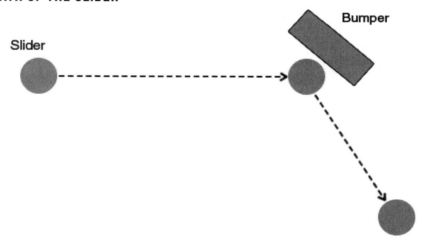

6. Teams will now test the effect of placing bumpers in a slider's path. Each team should test three different bumper positions one at a time, as follows:

    a. Select and diagram the bumper position as demonstrated.

    b. Predict what will happen when the slider collides with the bumper.

    c. Launch the slider so that it reflects off the bumper when they collide.

    d. Diagram the path of the slider.

    e. Conduct two more tests.

As the students explore, ask guiding questions like:

- What happens to the direction the slider is moving when a bumper is placed in its path?
- Which bumper position caused the greatest change in the slider's direction?

7. When all teams have finished testing, bring the class back together as a whole group. Ask teams to share and compare their findings.

8. As a class, make a general statement about what happens when an object moving in a straight line experiences an outside force. Write the statement on the board. Students will record the statement in their science notebooks. These questions will help guide students' thinking:

- How would you describe the path of the slider when there are no bumpers in the way?
- What happens when the slider runs into the bumper? Does this always happen?
- Why does the direction of the slider change when it runs into a bumper? (If students are having difficulty drawing the conclusion that the bumper applies a force to the slider, reread page 6 of Exploring Forces and Movement.)

Changing Force I

In this part of the explore phase students will investigate what happens when an increasing force is applied to the slider.

Advance preparation: For each team, cut a piece of string that is approximately 15 cm longer than the width of the desk. Securely tape a snack-size plastic bag to one end of the string, and tape the free end to the top of a slider (see Figure 9.3).

FIGURE 9.3. CHANGING FORCE I EXPERIMENTAL SETUP

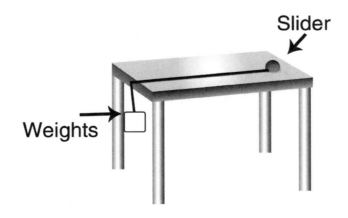

CLASSROOM CURLING:
EXPLORING FORCES AND MOTION

1. Draw the "Changing Force I" data table on the board.

2. Demonstrate the process students will use to test the effect that changing the force in the direction of the motion has on the speed of the slider. Put the slider near one edge of a desk and stretch the string across the desk so that the plastic bag hangs over the opposite side as illustrated in Figure 9.3. Put enough weight in the bag so that when it is released it slowly pulls the slider across the desk.

3. Repeat the demonstration using a stopwatch to time how long it takes the slider to reach the other side of the desk. Record the time in the data table that is on the board.

4. Give each group one slider with the string and plastic bag attached, a stopwatch, weights, and the "Changing Force I" data sheet.

5. Tell students to follow the procedure and the demonstration to complete their investigation. As the students explore, ask guiding questions like:

   • How does increasing the weight impact the time it takes for the slider to travel across the table?

   • What direction is the weight moving?

   • What direction is the slider moving?

   • What do you think would happen if the slider were heavier? (If desired, place a weight on top of the slider to test this.)

6. Students now find the mean (average) of the three trials for each test. Caution students to be sure to use the observed rather than the predicted times.

7. Ask students to create a bar graph with time on the vertical axis and number of weights on the horizontal axis. If they are not mathematically ready to calculate the mean, tell them to use the median (middle) time for their graph.

8. After creating the graph, students should share and compare with the whole class.

9. As a class, write a general statement about how increasing the force changes the motion of the slider. Post this statement on chart paper in a prominent place for the remainder of the unit.

Changing Force II

1. Draw the "Changing Force II" data table on the board.

2. Set up a ramp for the slider to travel down as shown in Figure 9.4 (p. 126). Adjust the angle of the ramp so that the slider will travel an additional 30 cm after it leaves the ramp.

FIGURE 9.4. CHANGING FORCE II RAMP SETUP

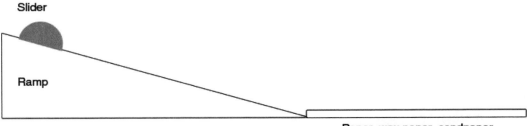

Slider

Ramp

Paper, wax paper, sandpaper

3. Demonstrate the process students will use to test the effect that friction has on the speed of the slider by releasing a slider from the top of the ramp. Allow it to continue in motion until it stops. Measure the distance traveled from the bottom of the ramp. Record the distance in the proper location on the data table. Place a piece of printer paper, wax paper, or sandpaper at the bottom of the ramp and repeat the demonstration.

4. Give each group materials for the ramp; a slider, one piece each of printer paper, wax paper, and sandpaper; and the "Changing Force II" data sheet.

5. Tell the students to set up their ramp as you demonstrated, adjusting the height until the slider travels about 30 cm after leaving the bottom of the ramp. After students have the ramp set up, they will do three trials without any paper, recording the distance traveled after each trial. They will then repeat the process for each of the three types of paper. Students will begin making predictions after a pattern in the data begins to emerge.
As the students explore, ask guiding questions like:

   - How do the different types of paper compare?
   - What do you notice about the way the slider moves on the different types of surfaces/ papers?
   - How does the texture of the paper affect the distance the slider travels?

6. Have students average the three trials for each test. Caution students to be sure to use the observed rather than the predicted times.

7. Have students create a bar graph with distance traveled on the vertical axis and surface/paper type on the horizontal axis. If they are not mathematically ready to calculate the average, tell them to use the middle distance for their graph.

8. After creating the graph, students should share and compare with the whole class. The comparison should include discussion about how the texture of the paper affected the distance the slider traveled. Students should make the observation that the slider traveled shorter distances on rougher paper.

9. Tell students that this is due to friction, and explain that they will be using several books to learn about friction. Hold up *Exploring Forces and Movement*, *Forces*, *Motion*, and *What Is a Force?* and ask students, "How can we find information about friction in these books? What text features will help us?" Guide students toward identifying the table of contents, index, and headings as text features that will be helpful to them. They will read what they find to confirm what they learned through their firsthand exploration. Students will record any new information in their science notebook. Students will now use the "Defining Friction" graphic organizer to develop a definition for friction based on their firsthand observations and the texts.

10. As a class, write a general statement about how increasing the force of friction impacts the speed of the moving object. Post this statement on chart paper in a prominent place for the remainder of the unit.

Down It Goes

In this part of the exploration, students will identify gravity as a force that pulls objects toward the Earth.

1. Lead a guided discussion in which students reflect on several aspects of the investigations they have just completed. These guiding questions will help students focus on the topic:

   • What direction were the weights going in "Changing Force I"? (Students should respond that the weights were going down.)

   • What had to be happening for the weights to move in a downward direction? (Students should respond that a force had to be acting on the weights.)

2. Crumple a piece of paper and hold it out with an outstretched arm. Ask students to predict what will happen when you let go of the paper. Students will most likely predict that the paper will drop to the ground. Release the paper, acknowledge that the class prediction was correct, and ask if the same force that pulled the weights down also pulled the paper down. Discuss briefly.

3. Demonstrate a slider moving down a ramp. Begin with the ramp in a flat position with a slider at one end. Ask students to predict what will happen when you lift the end of the ramp. Since they have experimented with ramps and sliders, they will know what to expect. Slowly elevate the end of the ramp with the slider on it. Stop when the slider begins to move down the ramp. Ask the students the following or similar questions to focus the students' attention on the force that pulls the slider down the ramp.

- What direction is the slider moving?

- How does the direction of the slider compare with the direction the weights and the crumpled-up paper were moving? (Students should respond that all are moving in a downward direction.)

- Something must be pulling on these objects, otherwise they would not move. Do you see anything pulling on these objects? How can we determine what is pulling on the weights, the paper, and the slider?

4. Tell students that they are going to use texts to find out what is pulling the objects downward. Students will use various text features in the books *Exploring Forces and Movement*, *Forces*, *Motion*, and *What Is a Force?* to locate information about gravity. They will read what they find to confirm what they learned through their firsthand exploration. Students will record any new information in their science notebook. Students will now use the "Defining Gravity" graphic organizer to develop a definition for gravity based on their firsthand observations and the texts.

Assess this phase: Again, use formative assessment to determine if students are developing appropriate concepts around forces, motion, friction, and gravity. Observe students during each of the explorations and use guiding questions to focus student attention and encourage them to dig more deeply into the explorations. If students struggle to compose general statements after each exploration, help them make connections by demonstrating key aspects of the activities while asking clarifying questions that assist the students in connecting the evidence they have gathered with the content they are learning. Student-generated definitions for friction and gravity provide evidence of their understanding of these concepts, as well as their ability to synthesize information from the hands-on experiences and the texts. If students struggle to complete this, provide support for individual students or complete the "Defining Friction" and "Defining Gravity" graphic organizers as a teacher-led, whole-class activity.

EXPLAIN

In the explain phase students will apply what they have learned about forces and motion to explain the activities of the boy in *Newton and Me*. They will then answer the question, *What happens to the motion of a slider when different forces are applied?*

1. Before reading *Newton and Me,* tell students that you would like for them to connect what is happening in the book to the explorations they just completed. If students are unfamiliar with the connecting-text-to-self reading strategy, conduct a mini-lesson on text-to-self connections. Then model an example of a text-to-self connection as you start reading. If students are familiar with this strategy, instruct the students to raise their hand when something you have read reminds them of something from their explorations, and ask a few students to share connections

with the class. Students may, for example, share that the slider moved farther on the wax paper compared with the sandpaper when you read page 5. After you have read the book, ask students to complete the "*Newton and Me* Connections" graphic organizer.

Making Connections

Making connections is a reading comprehension strategy in which readers draw on their prior knowledge and experiences to relate to a text. When a reader makes connections to a text, he or she is engaged in the reading process and is actively thinking about what he or she is reading. Making connections can thus greatly enhance comprehension. Connections are classified into three categories: text-to-self, text-to-text, and text-to-world. *Text-to-self connections* are those that link the reader's personal experiences to what is being read. *Text-to-text connections* are those that link information from the current text to one that was previously read. *Text-to-world connections* link the text to phenomena that occur in the world at large, but are not ones that the reader has necessarily experienced personally. In this activity, we invite students to make text-to-self connections between *Newton and Me* and what they have observed in the hands-on activities. We have slightly modified the definition of a text-to-self connection to suit our purpose. In this case, students make connections to a very specific set of experiences: the evidence from the hands-on activities.

---

2. Students will now prepare an explanatory report to answer the question, *What happens to the motion of a slider when different forces are applied?* Provide students with the following prompt to help them get started.

Writing Your Explanatory Report

Answer the question, *What happens to the motion of a slider when different forces are applied?* in an explanatory report. Be sure to include an explanation of how the slider

Starts and stops moving
Changes direction
Responds to friction
Responds to gravity

Use information from your explorations and reading as the foundation for your report. The "Putting the Pieces Together Report Planner" will help you organize your information.

Students will likely need guidance for writing the report. Suggest that students write a sentence or paragraph for each question on the report planner, depending on the level and writing ability of the students. Discuss the role that diagrams might play in the report and how they might be used to clarify written explanations. Refer students to the diagrams they created in the explore phase, and allow them to include these diagrams in their report (or re-create them if desired). Allow students to access any of the materials used throughout the inquiry as they write, and provide individual support to students as needed.

Assess this phase: The *Newton and Me* graphic organizer serves as the final formative assessment for the unit. Students should be able to make clear connections between the text and the hands-on experiences from the explore phase. If the class does not easily identify connections during the read-aloud or if students struggle to complete the graphic organizer, use guiding questions like "What part of the book does this remind you of?" or "Is this observation the same as what we read in the book?" to help students make the connections. The explanatory report (and the tip guide created in the expand phase) serves as summative assessments. Students should have clearly explained how the slider moves and changes direction as the result of a force being applied, and how the forces of friction and gravity affect the slider's motion. The "Science and Literacy Rubric" in Appendix 2 (p. 276) can be used to assess student performance, and the "Achievement Grading Standards" can translate this into a numerical grade. If students fall below 75%, return to the explore phase for additional work before returning to the writing prompt. Consider using the "Putting the Pieces Together Report Planner" as an alternative format for the report.

EXPAND

In the expand phase of the inquiry students will apply what they have learned about forces and motion to write and illustrate a tip guide for playing classroom curling. The guide should provide hints for success in a game of classroom curling. This can be done individually, in teams, or as a whole class. Students may want to test their tips to check for accuracy, or they can trade guides and test each other's work as a form of peer assessment. If students test their peers' work, provide time for revision before the tip guides are turned in. Consider using web-based publication applications to publish the guide online.

Assess this phase: This tip guide (along with the explanatory text produced in the explain phase) serves as summative assessment for this inquiry unit. Students should have clear tips that are explicitly based on the knowledge gained from the unit. The "Science and Literacy Rubric" in Appendix 2 can be used to assess student performance, and the "Achievement Grading Standards" (also in Appendix 2, p. 277) can translate this into a numerical grade. If students fall below 75%, return to the explore phase for additional work before returning to the creation of the tip guide.

REFERENCE

Driver, R., A. Squires, P. Rushworth, and V. Wood-Robinson. 1994. *Making sense of secondary science: Research into children's ideas.* New York: Routledge.

Name_____ Date_____

CHANGING FORCE I

Materials

- Weights
- Slider with weight bag attached
- Stopwatch or watch with second hand

Procedure

1. Add enough weight to the bag so that it slowly pulls the slider across the desk.
2. Record the number of weights in the correct column on the data table.
3. Conduct three trials and record the time it takes for the slider to be pulled across the desk for each trial.
4. Add more weight to the bag and repeat the process.

Test	Number of Weights	Time		
		Trial 1	Trial 2	Trial 3
1				
2				

What do you notice about the time it takes for the slider to be pulled across the desk when the weight is increased? _____

_____

_____

_____

_____

CLASSROOM CURLING:
EXPLORING FORCES AND MOTION

Name_____ Date_____

Based on what you noticed, predict how long it will take for the slider to be pulled across the desk when more weight is added.

Test 3 _____ number of weights   _____ predicted time

Test 4 _____ number of weights   _____ predicted time

Test	Number of Weights	Time		
		Trial 1	Trial 2	Trial 3
3				
4				

The weights are a force pulling on the slider. How does increasing the weight (force)

change the motion of the slider? _____

_____

_____

_____

_____

_____

_____

_____

CLASSROOM CURLING:
EXPLORING FORCES AND MOTION

Name_____ Date_____

CHANGING FORCE II

Materials

- Slider
- Ramp
- Printer paper, wax paper, sandpaper
- Ruler

Slider

Ramp

Paper, wax paper, sandpaper

Procedure

1. Adjust the height of the ramp so that the slider travels about 30 cm beyond the bottom of the ramp.
2. Release the slider from the top of the ramp. Measure the distance it travels beyond the bottom of the ramp. Conduct three trials.
3. Place a piece of wax paper at the bottom of the ramp. Repeat the procedure.

Test	Surface/ Paper Type	Distance Traveled		
		Trial 1	Trial 2	Trial 3
1				
2				

What do you notice about the distance the slider travels beyond the bottom of the ramp when the surface is changed? _____

_____

_____

_____

CLASSROOM CURLING:
EXPLORING FORCES AND MOTION

Name_____ Date_____

Based on what you noticed, predict what will happen for the other types of paper.

Test 3 _____ printer paper   _____ predicted distance

Test 4 _____ sandpaper   _____ predicted distance

Test	Surface/ Paper Type	Distance Traveled		
		Trial 1	Trial 2	Trial 3
3				
4				

How does changing the texture (smoothness or roughness) of the paper change

the motion of the slider? _____

_____

_____

_____

_____

_____

_____

Name_____ Date_____

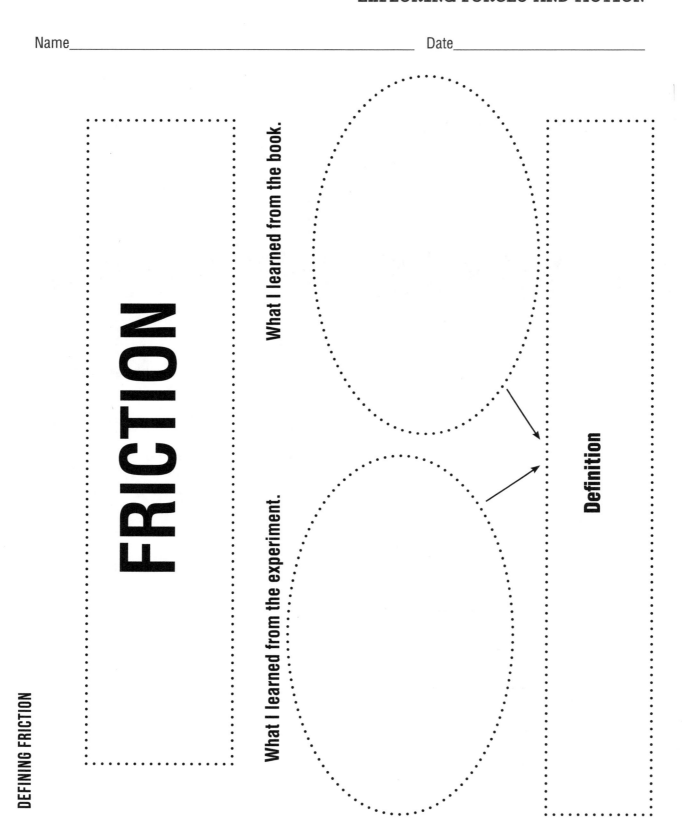

FRICTION

What I learned from the book.

What I learned from the experiment.

Definition

DEFINING FRICTION

Name_____ Date_____

DEFINING GRAVITY

GRAVITY

What I learned from the book.

What I learned from the experiment.

Definition

CLASSROOM CURLING:
EXPLORING FORCES AND MOTION

Name_____ Date_____

NEWTON AND ME CONNECTIONS

Read the quotes from *Newton and Me* on the left side. Write what the quote reminds you of from your explorations, or your text-to-self connection, on the right side.

Newton and Me	Text-to-Self Connection
"The ball won't roll far in the rough, grassy yard. It rolls much farther on a surface that is smooth and hard"	
"I put down the truck on ground that was flat. Until I would push, my truck stayed where it sat."	
"Going downhill my truck really speeds."	
"But going downhill, I needed to slow down. I did that by dragging my feet on the ground."	

CLASSROOM CURLING:
EXPLORING FORCES AND MOTION

Name_____ Date_____

PUTTING THE PIECES TOGETHER REPORT PLANNER

Answer each question in the space provided. Then use the information to answer the central question, *What happens to the motion of a slider when different forces are applied?*

How does the slider start and stop moving?	**What causes the slider to change direction?**
How does friction affect the motion of the slider?	**How does gravity affect the motion of the slider?**

What happens to the motion of a slider when different forces are applied?

Chapter 10

Beaks and Biomes: Understanding Adaptation in Migrating Organisms

OVERVIEW

In this unit, students conduct direct and indirect investigations to answer the question, *How is the sanderling adapted to survive in two different environments*? The unit focuses on the sanderling, a shorebird that migrates from warm, sunny beaches to the Arctic tundra to breed each year. The direct investigation in this lesson involves students in simulations of the sanderlings' feeding habits. In the indirect investigation, students take part in idea circles in which they work in collaborative groups to learn about the tundra through informational texts.

This unit uses an electronic book to set the stage for learning and build students' interest in sanderlings. The book is used again at the end of the lesson to confirm students' knowledge and paint a complete picture of the sanderlings' complex life history. The unit also includes a RAFT (**R**ole, **A**udience, **F**ormat, **T**opic) prompt as a means of assessing student understanding. More information about idea circles and RAFT prompts can be found in Appendix 1.

OBJECTIVES

- Explain the concept of adaptation as it relates to the sanderling
- Describe the shore and tundra environments
- Practice the following science process skills: predicting, observing, and collecting and analyzing evidence
- Read and discuss informational text to locate information
- Write nonfiction text in journal format

STANDARDS ALIGNMENT

National Science Education Standards
SCIENCE AS INQUIRY

- K–4, 5–8 Abilities Necessary to Do Scientific Inquiry

LIFE SCIENCE

- K–4 Characteristics of Organisms
- 5–8 Diversity and Adaptations of Organisms

Common Core State Standards for English Language Arts
WRITING

- Grades 3–5 Text Types and Purposes
- Grades 3–5 Production and Distribution of Writing
- Grades 3–5 Research to Build and Present Knowledge

For a detailed standards alignment, see Appendix 3 (p. 282).

TIME FRAME

- Six 45-minute class periods

SCIENTIFIC BACKGROUND INFORMATION

The sanderling, a small sandpiper, is one of the most widespread wintering shorebirds in the world. Sanderlings are typically 7.1–7.9 in. (18–20 cm) in length and weigh 1.4–3.5 oz. (40–100 g). They are distinguished from other species of sandpipers by their pale nonbreeding plumage, white face, black bill and legs, and a broad white wing stripe, bordered in black, which is visible in flight. Sanderlings have a long, thin, and pointed beak, which is used to probe for food.

During the winter months, the birds live on temperate and tropical sandy beaches along the coasts of North and South America. They feed on aquatic and terrestrial invertebrates such as mole crabs. Sanderlings can be seen running along the water's edge, darting in and out of the waves as they feed.

Beginning in March, adult birds migrate to the Arctic tundra to breed during the summer months, traveling thousands of miles to a vastly different environment. Instead of warm temperatures and sandy beaches, they find cooler temperatures and land covered by small shrubs, grasses, and mosses. Rather than mole crabs, the birds feed on insects and some plant material. The sanderlings build nests on the ground, each of which typically holds three to four dull green, spotted eggs. Eggs are incubated for approximately 24 days. Within 12 hours of hatching, the chicks leave the nest to join the adults in feeding. The chicks begin to fly within two weeks of hatching. The adult sanderlings begin their southward migration in July and early August, but the juveniles do not leave the breeding grounds until August and September. After the juveniles migrate south, they remain at the winter grounds for at least a year.

Learn more about the sanderling, listen to its call, and watch video at The Cornell Lab of Ornithology's All About Birds website: *www.allaboutbirds.org/guide/Sanderling/id* (QR Code 1).

QR Code 1

MISCONCEPTIONS

The concept of adaptation is difficult, and one about which students may have scientifically inaccurate preconceived notions, or misconceptions. Table 10.1 lists some documented misconceptions about adaptations along with a brief scientifically accurate concept.

TABLE 10.1 COMMON MISCONCEPTIONS ABOUT ADAPTATION

Common Misconception	Scientifically Accurate Concept
Organisms choose to adapt to their environments.	Adaptations are inherited traits that are passed from parent to offspring. They enhance an organism's ability to survive and reproduce.
Adaptations are acquired during the lifetime of the organism.	
Organisms adapt to changing environments. These adaptations are then inherited.	

Pay careful attention to the language students use in their conversations and writing throughout the lesson. Language that suggests that adaptations are the result of an organism consciously selecting one adaptation over another may indicate that the student does not understand that adaptations are inherited traits that enhance survival or reproduction. Language that suggests that an adaptation occurs within an individual organism's lifetime is another indicator that the student does not understand that adaptations are passed from generation to generation. When it appears that students have these misconceptions, provide additional evidence through hands-on investigations or informational texts that supports the accurate concept.

TEXT SET

Beaks! by Sneed B. Collard III (Watertown, MA: Charlesbridge, 2002); explanation, Flesch-Kincaid reading level 4.2.

Bird beaks come in a variety of shapes and sizes, depending on the type of food the bird eats and the manner in which it gathers or hunts its food. Learn about the many varieties of beaks and test your "beak-ability" by matching birds' beaks to their foods. Colorful painted cut-paper illustrations vividly depict bird species and their food sources.

QR Code 2

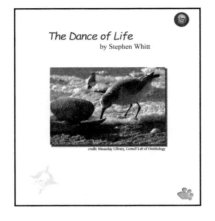

The Dance of Life by Stephen Whitt (Columbus: Ohio State University, 2009); narrative expository, Flesch-Kincaid reading level 5.1. *http://bit.ly/th6TrM* (QR Code 2).

This nonfiction text describes the life cycle, migration, and behavior of the sanderling. The article is available at three grade levels: K–1 (reading level 1.8), 2–3 (reading level 3.1), and 4–5 (reading level 5.1). Each grade level is available in three formats: a text-only document (PDF file), a full-color illustrated book (PDF file), and electronic books with recorded audio narration.

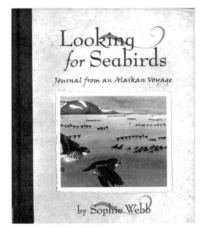

Looking for Seabirds: Journal From an Alaskan Voyage by Sophie Webb (Boston: Houghton Mifflin, 2004); journal, Flesch-Kincaid reading level 7.4.

Author and artist Sophie Webb spent a month working as a seabird observer along the Aleutian Island chain in Alaska. Experience her journey through her detailed writing and colorful, lifelike watercolor illustrations. Webb authored a similar book, *My Season With Penguins*, which details a research expedition in Antarctica.

Red Knot: A Shorebird's Incredible Journey by Nancy Carol Willis (Middletown, DE: Birdsong Books, 2006); diary (journal), Flesch-Kincaid reading level 4.4.

Follow the 20,000 mile annual migration of a shorebird called a red knot as it travels from the tip of South America to the Arctic tundra and back. Short dated entries complement large colorful illustrations. Readers will notice many similarities between the red knot and the sanderling.

MATERIALS

Per Student

- 1 pair of tweezers of an appropriate size to pick up the smaller items in the environment
- Copies of the Venn diagram graphic organizer and RAFT writing prompt

Per Small Group

- Two 8" square aluminum pans
- Walnut halves
- Sand
- Spanish moss (available at craft stores)
- A variety of stones and pebbles
- A variety of shells
- Items to represent insects and larvae (e.g., pom-poms, wooden beads or buttons, foam insects)
- Larger items to represent lemmings and other small rodents (e.g., wooden spools)
- Copies of the "Idea Circle Graphic Organizer"

Per Class

- Chart paper and markers
- A set of books about the tundra (at least one per student)
- Computer with internet access and speakers
- Bird field guides
- Interactive whiteboard (optional)
- Copies of supporting documents

SUPPORTING DOCUMENTS

- "Idea Circle Graphic Organizer"
- Venn diagram
- RAFT writing prompt

SAFETY CONSIDERATIONS

Make sure that students do not have food allergies before using food items in this lesson. Instruct students to not eat any of the food items used in the lab. Demonstrate how to use the tweezers to mimic the sanderling's feeding behavior. Monitor students to ensure that they are using the forceps appropriately.

Beaks and Biomes Inquiry Unit

ENGAGE

In the engage phase, students are introduced to the sanderling and its life history and migratory patterns.

1. Engage students by playing the first six pages of the grades 4–5 electronic book *The Dance of Life* as a read-aloud. The electronic book works well with an interactive whiteboard, or it could be played on a computer equipped with speakers. Alternatively students might read and listen to the first six pages at individual computers or at a listening center. Paper versions of *The Dance of Life* are also available (text-only and illustrated book) if computers are not available. The paper versions could be printed, copied, and distributed to individual students.

2. Discuss the first six pages with students as they read, focusing on understanding the vocabulary terms used as well as the images and diagrams. Ask students to describe the shape of the sanderling's beak: long, thin, and pointed. Ask students to think-pair-share* about why the shape of the sanderling's beak might be important. Allow a few pairs to share their thinking with the class.

3. Read *Beaks!* aloud, focusing on the relationship between each bird's beak and its diet. The bright and colorful illustrations invite students to carefully observe the shape of each bird's beak. Allow students to freely share observations and ideas, but keep them focused on the purpose for reading the text with guiding questions such as:

   - What did you notice about the birds' beaks and the types of food each bird eats?
   - How do birds with long pointy beaks get their food?
   - What types of foods do birds with short strong beaks eat?

   Listen for answers focusing on how the beak seems to work well with the type of food the bird eats or the method the bird uses to obtain food. When students come to the realization that the shape of the beak determines how it functions, introduce the term *adaptation.* Work with the students to develop a definition for this term, and post it in a prominent location.

4. Return to *The Dance of Life* and reread pages 5 and 6. Ask students to think-pair-share about how the sanderling might be adapted to two very different environments. This discussion serves as a transition into an exploration of the sanderling's diverse habitats.

* Think-pair-share is a three-step process that begins with students independently thinking about the prompt. Students then pair up to discuss their thinking and identify the best response. The process concludes with each pair sharing their thinking with the whole group.

Assess this phase: At this point, only formative assessment is needed. Monitor student comprehension of *The Dance of Life* during the class discussion. Students should take away three main ideas from the first six pages: the shore habitat of the sanderlings, mole crabs as a food source, and that the sanderlings migrate to the Arctic each year. If students struggle with these concepts, consider rereading with additional emphasis on these points or adding supplemental videos of sanderlings at the beach. You might also use the modified versions of *The Dance of Life*, if appropriate.

After listening to *Beaks!* students should make the connection between the shape of a bird's beak and the food it eats. If students are having trouble with this concept, you might consider demonstrating a variety of tools that represent beak shapes and how they obtain food. An aquarium net or slotted spoon can represent a beak that filters food from the water, a straw represents a beak (like a hummingbird's) that sips, and a nutcracker represents a beak that cracks open the hard shells of nuts or seeds. Further conversation around these tools will help students master the needed concepts before moving on to the explore phase.

EXPLORE

In the explore phase, students investigate the two environments, comparing their characteristics and identifying potential food sources for the sanderlings. Begin with the more familiar of the two environments: the beaches that serve as the wintering grounds for the birds.

Advance preparation: For the shore feast, partially fill 8″ square aluminum cake pans with sand (one pan per small group) and add a variety of items to represent food for the sanderling. We used walnut halves to represent mole crabs, shells to represent various mollusks, wooden beads to represent insects, and pebbles and rocks of various sizes (which simply represent rocks typically found in a shore environment). Prepare a key so that students will know what each element represents.

For the tundra feast, partially fill 8″ square aluminum cake pans with Spanish moss (one pan per small group) and add a variety of items to represent food for the sanderling. We used pom-poms, wooden buttons, foam insects, and gummy worms to represent insects and larvae, and wooden spools to represent lemmings and other rodents. We also included pebbles and rocks of various sizes. Prepare a key so that students will know what each element represents.

Idea Circle

For the idea circle, gather a wide variety of reference books about the tundra. The books should represent varied reading levels and formats; see a suggested list of books below. This list supports both reader choice and differentiation. Give a copy of the "Idea Circle Graphic Organizer" to each group.

Suggested Tundra Books
Add or substitute additional titles as necessary. Be sure to include a variety of reading levels and formats to allow for student choice.

- *Arctic Appetizers: Studying Food Webs in the Arctic* by Gwendolyn Hooks (Vero Beach, FL: Rourke, 2009); reference, Flesch-Kincaid reading level 5.8.

- *Discovering the Arctic Tundra* by Janey Levy (New York: Rosen, 2008); reference, Flesch-Kincaid reading level 5.5.

- *Explore the Tundra* by Linda Tagliaferro (Mankato, MN: Capstone Press, 2007); reference, Flesch-Kincaid reading level 5.1.

- *The Frozen Tundra* by Philip Johansson (Berkeley Heights, NJ: Enslow, 2004); reference, Flesch-Kincaid reading level 4.2.

- *Living in the Tundra* by Carol Baldwin (Chicago: Heinemann Library, 2004); reference, Flesch-Kincaid reading level 4.9.

- *Tundra* by Aaron Frisch (Mankato, MN: Creative Education, 2008); reference, Flesch-Kincaid reading level 2.4.

- *Tundra* by Colleen Sexton (Minneapolis, MN: Bellweather Media, 2009); reference, Flesch-Kincaid reading level 2.9.

- *A Tundra Food Chain: A Who-Eats-What Adventure in the Arctic* by Rebecca Hogue Wojahn and Donald Wojahn (Minneapolis, MN: Lerner Publications, 2009); reference, Flesch-Kincaid reading level 5.4.

- *A Walk in the Tundra* by Rebecca Johnson (Minneapolis, MN: Carolrhoda Books, 2001); reference, Flesch-Kincaid reading level 4.3.

Part I: Shore Investigation

1. Engage students in a whole-class discussion of the beach environment, using the following guiding questions:

    • Have you ever been to the beach?

    • What did you notice while you were there?

    • What was the weather like?

    • What kinds of earth materials (sand, rocks, soil) were present?

    • What kinds of plants and animals did you see?

Record student responses on chart paper that can be clearly displayed throughout the remainder of the lesson. If students have not personally been to the beach, they may have seen the beach on TV or in movies or read about it. If not, read an appropriate nonfiction book or provide images for students to observe as a means of building background knowledge.

2. Invite students to participate in a "shore feast" simulation. Before beginning the shore feast, hold up a pair of tweezers. Ask the class how the tweezers compare with the sanderling's beak. Model how the tweezers can be used to mimic the probing behavior of the sanderling as it feeds.

3. Show students one of the prepared cake pans and, using the key for reference, review what each item represents. Explain to students that they will pretend that they are sanderlings, using the tweezers to determine what types of food they are able to eat at the beach environment. This is also a good time to review guidelines for successful collaborative group work.

4. Students will then work in collaborative small groups to snatch up food items with their tweezers, perfecting their "feeding" technique. While small groups are working, walk around and discuss the activity with students, using the following guiding questions:

    • What food is the easiest for the sanderling to eat?

    • What food is the most difficult for the sanderling to eat?

    • How do the foods that the sanderling is able to eat compare with those it cannot eat?

    • How does the sanderling's beak help it capture its food?

    • Have you noticed anything about the sanderling's beak that keeps it from eating certain types of food?

    • How would you describe the sanderling's feeding behavior?

5. Once students have completed the shore feast, conduct a brief whole-class discussion to debrief, using the following guiding questions:

- Were you able to successfully find food using the tweezers?
- What kinds of food were the sanderlings able to eat?
- What kinds of food were they not able to eat?

Record student findings on chart paper.

Part II: Tundra Investigation

1. Return to the map of the sanderling's migration shown on page 5 of *The Dance of Life* electronic book. Ask students to first identify where the beach environments are located, and then reread the text on that page. Point out the Arctic on the map, and explain that the environment that the sanderlings migrate to is called the tundra. While most students will have some prior experience or knowledge of a beach environment, they will most likely not know as much about the tundra. Lead a class discussion about the tundra, using the following guiding questions:

   - What do you know about the tundra?
   - What might sanderlings eat there?
   - How can we learn more about the tundra?

2. Explain to students that they will be using reference books and working in collaborative groups to learn more about this environment. Introduce the idea circle activity to the students. Explain that in the idea circle, they will read and discuss a variety of books in small groups to answer the following questions: What is the tundra like? What would you find there? What might sanderlings eat?

3. Begin the idea circle by organizing students into small groups. Each group member will then select and read a different text. If students have difficulty selecting appropriate texts for themselves, you may want to review the "five finger rule" first. To apply this rule, students should open a text to a random page and read it, holding up a finger each time they encounter an unknown word. If they have five fingers up before the end of the page, the book is most likely too difficult. Some students might need to select from a limited number of choices that are most appropriate for their reading level. Ideally, titles are not repeated within a small group, although they may be necessarily repeated within the class as whole. It is not necessary for students to take notes at this time, although some might prefer to do so. Consider allowing students to use slips of paper or sticky notes to bookmark pages with helpful information.

4. Students should read their books individually. When all students have finished reading their books, they should gather in their small groups and share what they've learned. Each group should complete the "Idea Circle Graphic Organizer," which provides spaces for students to record information along with the contributing text and group member.

5. Conclude the idea circle with a whole-class discussion in which groups share their findings. Record this information on chart paper and post it in a prominent location.

6. Now that students have learned about the tundra environment and made predictions about what the sanderlings might eat, it is time for a tundra feast. Just as in the shore feast, students work in cooperative groups to "feast" using the tweezers. A whole-class discussion at the conclusion of the activity confirms that the sanderlings were able to eat the insects. Student findings can be recorded on the chart paper begun with the idea circle.

7. Return to *The Dance of Life*. Finish reading and listening to the electronic book, using questions like "How does the text compare with what you learned from the tundra feast and the idea circle?" to help students link their hands-on experiences with the information presented in the text. Record student ideas on chart paper.

Assess this phase: Again, only formative assessment is needed. Monitor students' involvement during the shore and tundra feasts. Ensure that they are using the tweezers to properly mimic the sanderling's beak and that they are correctly identifying food sources in both the shore and tundra environments. During the idea circle, observe students as they read to ensure that all have selected an appropriate text. As they share information with their small groups, circulate and listen to their conversations. Ask guiding questions to help students extend their comprehension. Provide support to individual students and groups if they have difficulty locating information about the tundra or completing the graphic organizer. Finally, monitor students' comprehension of the remainder of *The Dance of Life*. Student responses to your questions will help you determine if they are able to link their hands-on experiences to the text. It is essential that students have a clear understanding of the two environments and the sanderlings' food sources in each before moving on to the explain phase.

EXPLAIN

In the explain phase, students draw from their experiences with the feasts and the texts to explain how the sanderling's adaptations allow it to survive in both environments.

1. Have students complete the Venn diagram (or other graphic organizer) to compare and contrast the sanderling's behavior on the beach and the tundra. Encourage students to refer back to the texts and feasts as needed.

2. Once students have completed the Venn diagram, they are ready to share their understanding

in a more formal way. Present the students with the writing prompt. This writing prompt is written as a RAFT prompt. RAFT stands for Role, Audience, Format, and Topic—four needed components of any effective writing prompt. We've found that the specificity found in RAFT prompts support students as they communicate their understanding of a new topic. We've also found that using boldfaced type to present the elements of a RAFT help students focus on the necessary components in their writing.

> You are a **scientist** who studies sanderlings by observing them in the wild. Write a **journal** for other **scientists** that describes your **observations of the behaviors that help sanderlings survive in these two different environments**. Include at least two entries.

3. Read over the prompt with students, answering questions as needed. This is a good time for a mini-lesson on the format of a diary or journal. Read excerpts from *Looking for Seabirds* and *Red Knot*, explicitly discussing the characteristics the texts share (dated entries that are presented in chronological order). Also discuss the use of first-person narration in *Looking for Seabirds*. The books should be readily available for student use during their work on the RAFT prompt. The books serve as mentor texts, supporting students as they write in the journal format. Encourage students to include illustrations with their journal entries.

Assess this phase: The Venn diagram is meant as a pre-writing tool, and should only be used for formative assessment. We recommend taking the time to collect and review the completed Venn diagrams to ensure that students have developed sufficient understanding and will be successful with the RAFT writing prompt. At this point, students should be able to explain that the sanderling migrates from the beach to the tundra and that while it eats different types of food in each place, it uses its beak to obtain food in each. If students struggle to complete the Venn diagram, consider returning to the hands-on activities and idea circle texts in a teacher-led activity.

Student responses to the RAFT writing prompt serves as summative assessment for this inquiry unit. Student responses should include the same information as the Venn diagram. The "Science and Literacy Rubric" in Appendix 2 (p. 276) can be used to assess student performance, and the "Achievement Grading Standards" (also in Appendix 2, p. 277) can translate this into a numerical grade. If students fall below 75%, return to the explore phase for additional work before returning to the writing prompt.

EXPAND

In the expand phase, students will further their understanding by investigating the adaptations of local birds. It is not necessary to be a birding expert to engage students in this activity. The goal is for students to apply what they have learned about adaptations to local wildlife. It is not necessary to identify the birds, but some students may want to know the identity of the birds they are observing. Have some field guides on hand for their use.

BEAKS AND BIOMES:
UNDERSTANDING ADAPTATION IN MIGRATING ORGANISMS

1. Begin by observing local birds. Students are likely to notice birds' physical characteristics first. Encourage them to pay attention to the birds' behaviors as well. Students should record their observations in a field journal or science notebook in words or in illustrations. They may observe feeding behaviors, nest building, flight patterns, and other behaviors and characteristics. Alternatively, students could observe birds in online videos or bird cams offered by zoos. Use these guiding questions to help students sharpen their observations:

   • What do you notice about the bird?

   • What does the bird appear to be doing?

   • What else is present in the bird's environment?

   • How does the adaptation you have observed help the bird survive?

2. Ask students to make predictions about what the bird will do next based on the pattern of behavior they have observed. After careful observation and data collection, the students might be able to draw some conclusions about the relationship between the birds' behavior and their physical characteristics. They may for example observe a bird repeating the pattern of flying out a short distance from a perch and then returning to the perch. From their observations they might draw the conclusion that the bird is catching flying insects.

3. To further enrich this study, you might elect to participate in eBird, a citizen scientist project run by The Cornell Lab of Ornithology. In the eBird program, students and adults across the country record bird sightings and submit their findings to the scientists at Cornell. Your bird counts help ornithologists study migration and other bird behaviors. Learn more at *http:// ebird.org/content/ebird* (QR Code 3).

QR Code 3

Assess this phase: At this point, assessment focuses on student observations of the birds' behavior. This is a time to assess science process skills: predicting, observing, and collecting data. Students are not only expanding their understanding of bird adaptations, but also developing their expertise with these process skills. Assessment should be focused not on a correct answer but on the process skills themselves. We recommend using the "Science Process Skills Rubric" in Appendix 2 to assess student abilities in this area.

BEAKS AND BIOMES:
UNDERSTANDING ADAPTATION IN MIGRATING ORGANISMS

Name_____ Date_____

IDEA CIRCLE GRAPHIC ORGANIZER

What We Learned About the Tundra	Where We Found It (title, team member)

Name_____ Date_____

VENN DIAGRAM

Compare and contrast the sanderling's behavior on the shore and in the tundra. How do its adaptations help it to survive in both places?

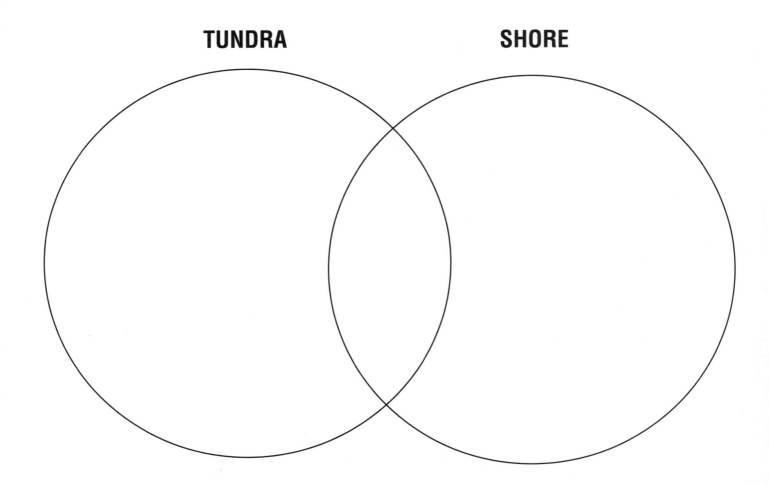

TUNDRA   **SHORE**

Name_____ Date_____

RAFT WRITING PROMPT

Use what you've learned from your investigations with the feasts and the texts to answer the following prompt.

You are a **scientist** who studies sanderlings by observing them in the wild. Write a **journal** for other **scientists** that describes your **observations of the behaviors that help sanderlings survive in these two different environments**.

_____

_____

_____

_____

_____

_____

_____

_____

_____

_____

Chapter 11
My Favorite Tree

OVERVIEW

Through this inquiry unit, the class works to answer the question, *What are the characteristics of trees?* Each group of students in the class selects a type of tree to investigate, such as oak, ash, or maple, and then answers the subquestion, "What are the characteristics of _____ trees?" The inquiry begins with a nature walk to observe trees. Students focus on the many different types of trees that can be found in a small area. Then, through research students discover that there are many different species of trees within a given genus (e.g., red, silver, and sugar maples).

Students practice their skills of observation while they pay attention to the details of the trees' characteristics. They analyze the differences between the trees and then synthesize the information to generate a list of common characteristics for their tree genus. They compile the information generated by each group into a guide to local trees. Students then apply what they have learned as they decide which trees are best suited for specific purposes in a park setting.

OBJECTIVES

- Recognize that trees are a diverse group of organisms
- Compare different species within a genus (e.g., sugar maple and silver maple)
- Write and illustrate informative text to convey information about trees
- Recognize that the characteristics of some trees make them well suited for specific uses
- Use multiple sources of information to verify facts

STANDARDS ALIGNMENT

National Science Education Standards
SCIENCE AS INQUIRY

- K–4, 5–8 Abilities Necessary to Do Scientific Inquiry
- K–4 Understanding About Scientific Inquiry

LIFE SCIENCE

- K–4 Characteristics of Organisms
- 5–8 Diversity and Adaptations of Organisms

Common Core State Standards for English Language Arts
INFORMATIONAL TEXT

- Grades 3–5 Craft and Structure

WRITING

- Grades 3–5 Text Types and Purposes

For a detailed standards alignment, see Appendix 3 (p. 282).

TIME FRAME

- Nine 45-minute class periods

SCIENTIFIC BACKGROUND INFORMATION

There are approximately 100,000 species of trees on Earth, accounting for 25% of all plants (Botanic Gardens Conservation International 2007). Within this number of species is an astonishing amount of diversity. Trees exhibit a wide variety of adaptations ranging from seeds that float to specialized leaves to root adaptations, and they are found on every continent but Antarctica.

A single genus of trees can include many different species. The largest genus of broadleaf trees, *Acer* (maple trees), includes at least 140 species (Chen 2007), 13 of which can be found in North America (U.S. Geological Survey 2006). Conversely, a single genus may also include only one species, as in the *Gingko biloba*.

Several characteristics can be used to differentiate one tree species from another. Leaves, bark, twigs, and fruit are commonly used to identify trees. Dichotomous keys organize information about these characteristics in a way that allows scientists to quickly navigate through large amounts of information and identify trees at the species level. Field guides are also used to differentiate between tree species, but these guides do not always include a dichotomous key. Dichotomous keys rely heavily on scientific terminology, whereas field guides include less terminology and often include drawings or photographs of many of the species included in the guide.

MISCONCEPTIONS

There are many misconceptions about plants, including misconceptions about photosynthesis, plant respiration, seed germination, and plant reproduction. Table 11.1 lists misconceptions reported by Barman et al. (2006) that are relevant to this unit.

Through this inquiry students begin to recognize this diversity through careful observation that leads to very basic classification. Children tend to have more trouble classifying plants than animals (Driver et al. 1994). This unit provides an opportunity to talk about trees as plants, not as a focus of inquiry, but in a casual way. It would be appropriate when reading *My Favorite Tree* to mention that trees are very interesting plants, or to mention that the trees are the biggest plants you see on the nature walk. Providing this scaffolding will help students begin to make the connection independently.

TABLE 11.1. COMMON MISCONCEPTIONS ABOUT TREES

Common Misconception	Scientifically Accurate Concept
Trees are not plants.	Trees are classified in the kingdom Plantae. They are multicellular organisms capable of producing their own food through photosynthesis. They generally contain chlorophyll, and their cellular structure includes components common to plants such as chloroplasts and a cell wall.
Plants have specific characteristics or parts such as flowers and stems, are green, and grow in soil.	Plants are a diverse group of organisms with adaptations that help enhance their survival. They are photosynthetic, sexually reproducing, and multicellular. Plants grow in a variety of conditions, some of which do not involve soil. Some plants, such as ferns, do not have stems or flowers. Others, such as mosses, do not produce seeds.

TEXT SET

My Favorite Tree: Terrific Trees of North America by Diane Iverson (Nevada City, CA: Dawn Publications, 1999); reference, Flesch-Kincaid reading level 4.3.

Each two-page spread of this engaging book introduces students to the wide variety of trees found in North America. The left-hand pages feature short comments by children and lifelike illustrations of their favorite tree, while right-hand pages provide encyclopedia-like entries about the tree.

National Audubon Society First Field Guide: Trees by Brian Cassie (New York: Scholastic, 1999); field guide, Flesch-Kincaid reading level undetermined.

Students will find more than 450 color photographs that capture the diversity of trees commonly found in North America. The text describing each tree is clear and concise.

Peterson First Guides: Trees by George A. Petrides (Boston: Houghton Mifflin Harcourt, 1998); field guide, Flesch-Kincaid reading level undetermined.

This compact field guide provides information about 243 trees common to North America. Illustrations of the twigs, fruit, and leaves are provided for each tree. The organization of the guide makes it easy to compare similar trees.

Trees, Leaves, and Bark by Diane Burns (New York: Gareth Stevens, 1998); field guide, Flesch-Kincaid reading level 3.8.

A guide to 15 trees found in the United States, providing information about their lifespan, uses, leaves, bark, and seeds.

MATERIALS

- Clipboards or science notebooks
- Paper and crayons for rubbings
- Computer with internet connection
- Digital camera (optional)
- Copies of supporting document

SUPPORTING DOCUMENT

- "Characteristics of My Tree" graphic organizer

SAFETY CONSIDERATIONS

Avoid areas where students could come in contact with poison ivy or similar plants. Show students pictures of poison ivy and similar plants and caution them to not touch anything that resembles the pictures.

My Favorite Tree Inquiry Unit

ENGAGE

In the engage phase, students begin to recognize that there are many different kinds of trees. Read aloud *My Favorite Tree* (left-hand pages only). As you are reading, comment on the many different types of trees. Wrap up with, "Wow! I didn't know there were so many different kinds of trees! How many types of trees do you think there are in our city (school, park, neighborhood)?" Allow students to make guesses about how many different types of trees might be found in the area.

Assess this phase: Formative assessment is used at this point in the inquiry to check student progress and help you determine if instructional modifications are needed. Student responses to your questions will help you gauge their prior knowledge of tree species and the types of trees found in the area.

EXPLORE

In the explore phase, students investigate tree diversity as they observe, identify, and research different species in a genus of trees.

1. Take students on a nature walk to observe trees. Bring along field guides such as *Trees, Leaves, and Bark* and *Peterson First Guide to Trees*, clipboards or science notebooks, and paper and crayons for rubbings. The walk can take place on school grounds, at a local park, through the neighborhood, at an arboretum, or any other place where a variety of trees grow. The emphasis is on observations, noticing, describing, and illustrating the unique characteristics of a tree in detail.

2. Start the walk by casually observing trees. Ask students what they notice about each tree. Encourage them to make careful observations, noticing the smallest details. Point out some tree characteristics like the shape of the leaves, the bark, fruits or seeds/nuts, and whether the tree is deciduous or evergreen. This is a good opportunity to point out the difference between simple and compound leaves. If the walk takes place somewhere that students are free to wander, you may need to regroup as a whole class from time to time to share general information about characteristics of trees.

3. Collect samples to bring back to the classroom, including leaves (or leaf rubbings), bark rubbings, and fruits and nuts. Digital photography can also be used to document trees and their leaves. If time permits, allow students to identify trees on site, recording a list of trees in their student science notebooks.

4. Return to the classroom and continue to identify trees using the leaves, rubbings, fruits and nuts, and photographs. Generate a list of trees identified by the class and post this on chart paper. Encourage students to observe and identify trees at home or around town and bring in rubbings and leaves. Continue to add to the list of trees posted in the classroom. When the list is sufficiently long, ask students to indicate which of the trees on the list is their favorite tree and why.

5. Divide students into small groups and assign each group a genus or type of tree that students have identified locally. For example, one group of students might be assigned oaks, a second maples, and a third pine. Consider using the students' favorite tree as an organizer and if possible, allow them to research that tree. Each group will use field guides and online resources to research the trees they have been assigned. Each group is answering the more specific question, "What are the characteristics of [insert tree type, e.g., maple] trees?" Each group will research three or four varieties within their tree's genus. For example, if they are researching maples, they might research a silver maple, sugar maple, and red maple. The goal of the research is to determine what distinguishes one tree from another within the genus, such as number of leaflets on leaf and shape of nut. Groups will complete a "Characteristics of My Tree" graphic organizer for each specific tree. They will then compare across the specific trees to generate a list of characteristics for the genus. An example of student work is shown at the end of the chapter.

A variety of field guides can be used, and several are mentioned in the text set. You can also contact your state's department of natural resources or forestry for information. Suggested online resources are listed below, with QR codes provided for your convenience. Use a scanning app on your smartphone, on your tablet, or with the webcam on your computer to scan and quickly access the online field guides.

- What Tree Is It? *www.oplin.org/tree* (QR Code 1). This describes trees in Ohio, but would work for much of the eastern or northeastern United States.
- Arborday.org Tree Guide, *www.arborday.org/treeguide* (QR Code 2)
- What Tree Is That? *www.arborday.org/trees/whattree/?TrackingID=908* (QR Code 3). Includes some technical language.
- Audubon Guides, *www.audubonguides.com/home.html* (QR Code 4). Requires signup. Reading level may be a bit high.

QR CODE 1 QR CODE 2 QR CODE 3 QR CODE 4

6. Develop a "Common Characteristics Banner" for their genus. The banner should be long enough to span the "Characteristics of My Tree" graphic organizers for their group.

Assess this phase: Formative assessment happens throughout this phase. Observe students on the nature walk and as they work to identify tree species. Ensure that they understand terminology and criteria used for classification (simple vs. compound leaves, leaflets, and so on) and that they are able to use print and online field guides. These skills are also important during the research phase. Make sure that students have selected sources that are accurate and age appropriate in terms of reading levels and sophistication of concepts. Monitor student work on the graphic organizers to ensure that students are focusing on distinguishing characteristics. Use guiding questions to help students look across species and identify similar characteristics within their genus.

EXPLAIN

In the explain phase, students demonstrate their expertise by sharing their findings and creating a local tree guide.

1. Each group will present their findings to the class. One group member will describe the common characteristics for their tree genus. Each group member will then describe the characteristics of a specific tree that distinguish it from the rest of the trees in that genus.

2. Have students organize the group products under a larger banner labeled "Trees," as shown in Figure 11.1.

FIGURE 11.1. ORGANIZATION OF GROUP PRODUCTS

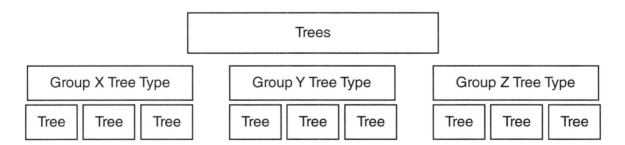

3. Create a class guide to local trees, with each student contributing a page. Decide as a class how each page will be formatted and what information will be provided. You may wish to have students use the field guides (from the explore phase) as mentor texts. Students can consider the information presented in the field guides and the format of the guide, then make decisions as a class about the information and format of their own guide. For example, they might include written descriptions, hand-drawn illustrations, photographs, and leaf and bark rubbings in the field guides.

Assess this phase: Student presentations serve as formative assessment. Each student's presentation should clearly describe distinguishing characteristics of their species, and the group should be able to explain how these species are related in one genus. If a group has difficulty identifying common characteristics, return to the research portion of the explore phase and use guiding questions to help the group look across species and focus on the relevant characteristics.

Student pages will serve as one form of summative assessment. Each page should describe a species of tree, its larger group, and its distinguishing characteristics. Illustrations and diagrams should be clearly labeled. The "Science and Literacy Rubric" in Appendix 2 (p. 276) can be used to assess student performance, and the "Achievement Grading Standards" (also in Appendix 2, p. 277) can translate this into a numerical grade. If students fall below 75%, return to the explore phase for additional work before revising their page for the class guide.

EXPAND

In the expand phase, students apply what they have learned to select trees for a local park.

1. Tell students that the class has been asked by a local park director to select the specific types of trees that will be planted in a new park. The director wants trees for each of the following park features:

    » Picnic area: Trees planted here must provide shade.

    » Wildlife area: Trees planted here must provide food and shelter for birds and squirrels.

    » Scenic walkways: Trees planted along the walkways should have showy flowers in the spring, have brightly colored leaves in the fall, or be very fragrant (e.g., pines). A combination of these types of trees would be best.

2. Reorganize so that each group is composed of students that researched different tree genera. A group might include a student who researched oaks, another who researched maples, and another who researched pines.

3. Working in their mixed groups, students will decide which trees would best fit the park director's needs. After selecting the trees, they will write and illustrate a paragraph describing the trees they have selected and explaining how these trees fit the park's needs. They will also create a map of the park indicating where the picnic area, wildlife area, scenic walkways, and trees will be located.

Assess this phase: The written product and map created in this phase also serve as summative assessment. Students should be able to apply the knowledge gained about their tree species to determine which trees would be best placed in each area. The "Science and Literacy Rubric" in Appendix 2 (p. 276) can be used to assess student performance, and the "Achievement Grading Standards" (also in Appendix 2, p. 277) can translate this into a numerical grade. If students fall below 75%, return to the explore phase for additional work to help students match their tree species to park features.

Investigating Different Types of Organisms

This inquiry unit can easily be adapted for the investigation of many different types of organisms. Follow the basic format of an authentic experience such as a nature walk for direct observations, identification and grouping using field guides or online resources, creation of a class guide, and application of new knowledge to a real-world situation. Here are some suggested text sets for birds, frogs, and insects.

TEXT SET

Birds

Olivia's Birds: Saving the Coast by Olivia Bouler (New York: Sterling, 2011); narrative expository, Flesch-Kincaid reading level 5.1.

Eleven-year-old Olivia Bouler, the writer and artist of this book, illustrates facts about birds with lovely, colorful drawings. Olivia describes how she sold drawings like these to raise over $150,000 for her Save the Gulf project after the devastating Gulf oil spill.

Thunder Birds: Nature's Flying Predators by Jim Arnosky (New York: Sterling, 2011); reference, Flesch-Kincaid reading level 5.9.

Well-known writer-artist Jim Arnosky describes different types of birds he observed and the features that make them unique. The many life-size paintings of birds are spectacular, including four foldout pages that can't help but garner a "Wow!"

You can also use various bird field guides, such as *Backyard Birds* by Karen Stray Nolting and Jonathan Latimer (New York: Houghton Mifflin, 1999) and *National Audubon Society First Field Guide: Birds* by Scott Weidensaul (New York: Scholastic, 1998).

Frogs

Frogs Sing Songs by Yvonne Winer (Watertown, MA: Charlesbridge, 2003); poetry, Flesch-Kincaid reading level 2.5.

Each lyrical four-line poem begins the same way: "Frogs sing their songs," and the poem goes on to describe how, when, where, or why they sing. Beautiful, realistic watercolor paintings show the frogs in their habitats, and thumbnail paintings illustrate a frog identification guide at the end of the book.

Frogs by Nic Bishop (New York: Scholastic, 2008); reference, Flesch-Kincaid reading level 4.8.

Straightforward text conveys basic information as well as cool and quirky facts about frogs. Large—*really large* in comparison to the real-life size of the frogs—color photos allow the reader to see details that would otherwise be very difficult to see.

You can also use various amphibian field guides, such as *Frogs and Toads* by Dave Showler and Barry Croucher (New York: St. Martin's Press, 2004) and *National Audubon Society First Field Guide: Amphibians* by Brian Cassie (New York: Scholastic, 1999).

Insects

Insect Detective by Steve Voake (Somerville, MA: Candlewick Press, 2010); narrative expository, Flesch-Kincaid reading level 4.6.

Beneath the ground, under rocks, in cracks in walls—you can see insects everywhere if you know where to look. Would-be insect detectives will find good information about specific insects as well as information about insects in general. Soft-color pen and watercolor drawings illustrate the text.

Insects In Danger by Kathyrn Smithyman and Bobbie Kalman (New York: Crabtree, 2006); reference, Flesch-Kincaid reading level 5.1.

This book describes the danger insects face from environmental circumstances that threaten their habitats.

You can also use various insect field guides, such as *National Audubon Society First Field Guide: Insects* by Christina Wilsdon (New York: Scholastic, 1998) and *Peterson First Guide to Insects of North America* by Christopher Leahy (New York: Houghton Mifflin Harcourt, 1998).

REFERENCES

Barman, C. R., M. Stein, S. McNair, and N. S. Barman. 2006. Students' ideas about plants and plant growth. *American Biology Teacher* 68 (2): 73–79.

Botanic Gardens Conservation International. 2007. Tree-BOL to barcode world's 100,000 trees. *www.bgci.org/worldwide/news/0463*

Chen, Y. S. 2007. Two newly recorded species of Acer (Aceraceae) in China. *Acta Phytotaxonomica Sinica* 45 (3): 337–340.

Driver, R., A. Squires, P. Rushworth, and V. Wood-Robinson. 1994. *Making sense of secondary science: Research into children's ideas*. New York: Routledge.

U.S. Geological Survey. 2006. Digital representations of tree species range maps from "Atlas of United States Trees" by Elbert L. Little Jr. (and other publications). *http://esp.cr.usgs.gov/data/atlas/little*

Name_____ Date_____

CHARACTERISTICS OF MY TREE

Tree Name:	
Leaf Shape	**Leaf Size**
Leaf Color (summer and fall)	**Fruit/Nut/Seed**

LEAF DRAWING

MY FAVORITE TREE

Name_____ Date_____

EXAMPLE GROUP PRODUCT

A team of three students is assigned maple trees as the tree type they will research. They decide to research the red, sugar, and silver maple. After looking at several resources each student completes a summary sheet for one of the trees. They then analyze the information looking for characteristics shared by all of the trees in the group.

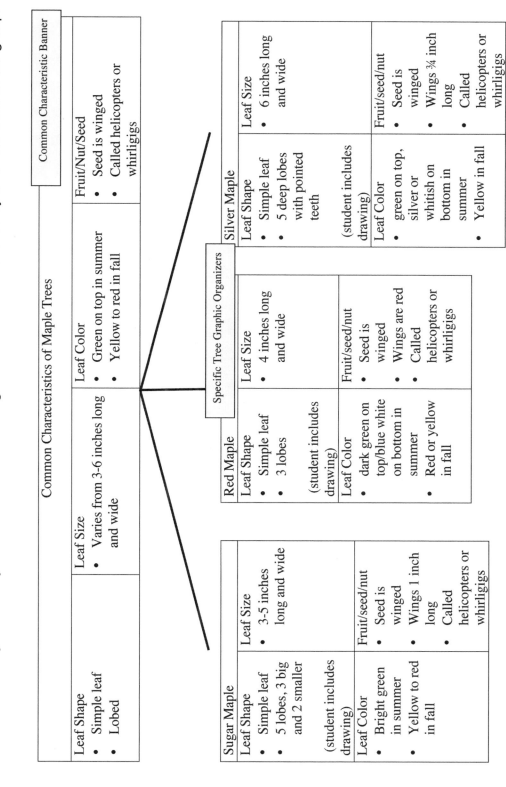

Common Characteristic Banner

Common Characteristics of Maple Trees

Leaf Shape	Leaf Size	Leaf Color	Fruit/Nut/Seed
• Simple leaf • Lobed	• Varies from 3–6 inches long and wide	• Green on top in summer • Yellow to red in fall	• Seed is winged • Called helicopters or whirligigs

Specific Tree Graphic Organizers

Sugar Maple

Leaf Shape	Leaf Size	Leaf Color	Fruit/seed/nut
• Simple leaf • 5 lobes, 3 big and 2 smaller (student includes drawing)	• 3–5 inches long and wide	• Bright green in summer • Yellow to red in fall	• Seed is winged • Wings 1 inch long • Called helicopters or whirligigs

Red Maple

Leaf Shape	Leaf Size	Leaf Color	Fruit/seed/nut
• Simple leaf • 3 lobes (student includes drawing)	• 4 inches long and wide	• dark green on top/blue white on bottom in summer • Red or yellow in fall	• Seed is winged • Wings are red • Called helicopters or whirligigs

Silver Maple

Leaf Shape	Leaf Size	Leaf Color	Fruit/seed/nut
• Simple leaf • 5 deep lobes with pointed teeth (student includes drawing)	• 6 inches long and wide	• green on top, silver or whitish on bottom in summer • Yellow in fall	• Seed is winged • Wings ¾ inch long • Called helicopters or whirligigs

Chapter 12
Come Fly With Me

OVERVIEW

In this inquiry, students work to answer the question, *How does the shape of a bird's wing affect its flight?* Students first investigate how factors such as wing length and shape affect the takeoff and flight of a paper airplane. They use this information to make predictions about birds' flight and then confirm or revise predictions by watching video clips. Finally, students apply these principles to flying machines such as gliders, helicopters, and airplanes.

A unique feature of this inquiry is the use of a directed viewing-thinking activity, or DVTA. This is a modification of a directed listening-thinking activity (DLTA), a research-based strategy for promoting comprehension of a text (Stauffer 1975). The DLTA strategy prompts students to make predictions about a text and then confirm or refute those predictions by listening to a read-aloud. We've created the DVTA to promote the development of science content knowledge by making predictions about an image, then using video clips to revise those predictions. In this way, the DVTA develops an underutilized literacy skill: the ability to critically view a diagram or video clip and obtain evidence from it. More information about the DVTA can be found in Appendix 1. Students also practice drawing diagrams, or infographics, and practice using the text features (captions, headings, labels) associated with them.

Finally, the unit also provides an opportunity to develop vocabulary about flight through the use of a word wall and kinesthetic activity. Students practice using the vocabulary with the help of sentence stems. Later, they incorporate the vocabulary into their own writing.

OBJECTIVES

- Explain the relationship between structure and function in birds' wings
- Practice the following science process skills: predicting, observing, and collecting and analyzing evidence
- Design and interpret infographics
- Write expository text to demonstrate understanding

STANDARDS ALIGNMENT

National Science Education Standards
SCIENCE AS INQUIRY

- K–4, 5–8 Abilities Necessary to Do Scientific Inquiry

LIFE SCIENCE

- K–4 Characteristics of Organisms
- 5–8 Structure and Function in Living Systems

Common Core State Standards for English Language Arts

INFORMATIONAL TEXT

- Grades 3–5 Craft and Structure
- Grades 3–5 Integration of Knowledge and Ideas

WRITING

- Grades 3–5 Text Types and Purposes

SPEAKING AND LISTENING

- Grades 3–5 Comprehension and Collaboration

For a detailed standards alignment, see Appendix 3 (see p. 282).

TIME FRAME

- Ten 45-minute class periods

SCIENTIFIC BACKGROUND INFORMATION

Birds and other flying animals depend on the principles of lift and thrust to be able to fly (Kazilek 2009). Lift occurs when the air pressure below an object is higher than the pressure above, causing the object to become airborne. Birds and other flying animals' wings are slightly curved. This shape forces air traveling over the top of the wing to travel farther (and faster) than the air below. The faster-moving air on top exerts less pressure than the slower-moving air below, and lift is generated.

However, lift alone won't allow a bird to start flying. Thrust is also needed. Thrust generates moving air and allows the bird to take off from the ground. Birds generate thrust by flapping their wings. Some birds run along the ground to gain speed, while others paddle across the surface of a body of water. Still others perch in trees and catch the wind from there. They also take off facing the wind whenever possible to gain extra momentum.

Birds have light bodies to facilitate flight. Their skeletons are made of light, hollow bones. They also have a very efficient heart and respiratory system and strong pectoral muscles for flapping wings (Deinlein n.d.). Several types of feathers, including flight, tail, and contour feathers, serve various purposes with respect to flight (Kazilek 2009).

Birds have a variety of wing shapes. Different wing shapes lend themselves to different types of flight. In general, scientists have identified four major types of wings (London 2011), as depicted in Figure 12.1.

FIGURE 12.1. FLIGHT SILHOUETTES

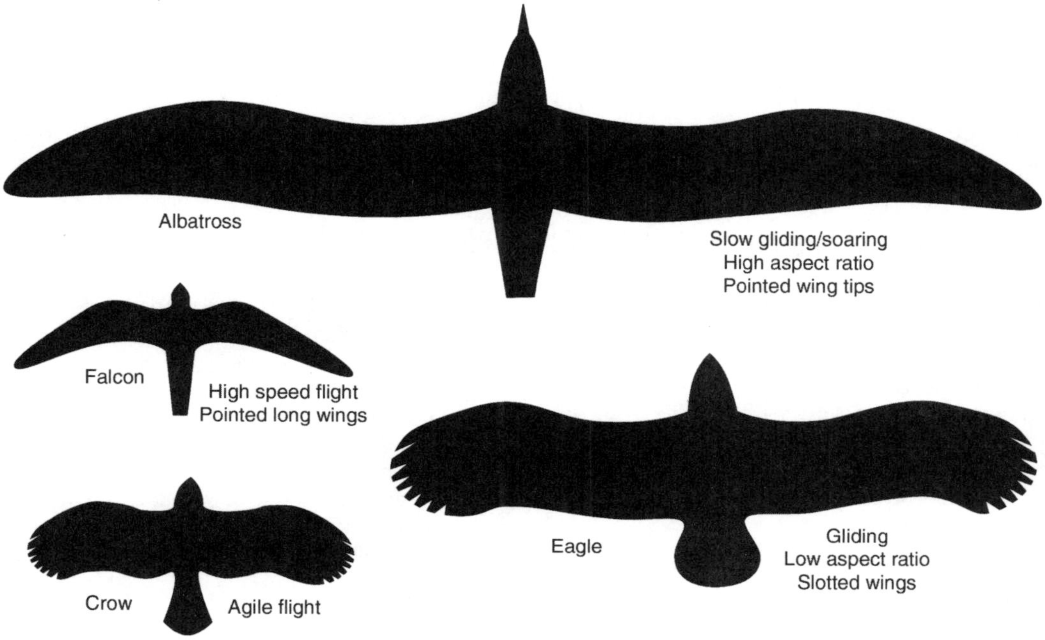

Source: Reprinted from L. Shyamal. December 2007. Wikimedia Commons. *http://enwikipedia.org/ wiki/File:FlightSilhouettes.svg*

Elliptical wings (short, rounded wings) are found on birds that live in forests or other heavily vegetated areas, such as pheasants and grouse. These birds can maneuver in tight spaces and take flight quickly to avoid predators. Short, pointed wings, such as those found on falcons and swifts, allow for fast flight. Long, wide wings, such as those found on a seagull or albatross, are best suited for slow flight, or for gliding and soaring. Other birds have long, wide wings with "slots," or gaps between the tips. These provide extra lift, allowing the bird to maintain flight even with the additional weight of prey. Birds like eagles and vultures have these soaring wings with deep slots. In this unit, we'll focus on just two of these four wing types: elliptical wings and long, wide wings.

MISCONCEPTIONS

There are many common misconceptions about bird characteristics, behaviors, and classification (Cardak 2009). Misconceptions related to bird adaptations that play a role in flight are listed in Table 12.1 (p. 176).

In this unit, students will investigate one of the physical characteristics of birds that play a role in flight, the shape of the wing. In the explore phase students are given an opportunity to share their thinking about how birds fly. If students' comments at that time reveal misconceptions, compare the students' thinking with the information presented in *Birds: Nature's Magnificent Flying Machines*. The goal is to help students reason through their misconceptions rather than correcting them directly. Their misconceptions may not be resolved at this time, but the students will have started questioning their thinking. As the unit progresses, students may express an interest in pursuing their misconceptions more thoroughly. This can be done through independent student research online or with print materials.

TABLE 12.1. COMMON MISCONCEPTIONS ABOUT BIRD ADAPTATIONS RELATED TO FLIGHT

Common Misconception	Scientifically Accurate Concept
Birds can fly because of the air spaces between their cells.	A variety of physiological features play a role in avian flight. These features include feathers, wings, fewer bones, pneumatization (or hollowing) of bones, large and efficient respiratory system, air sacs connected with air spaces in bones, large powerful heart, flight muscles with rich blood supplies, and a nervous system that facilitates quick responses (Ehrlich, Dobkin, and Wheye 1988).
The air sacs that help birds fly are located in their feet.	
Birds can fly because they are light animals.	

TEXT SET

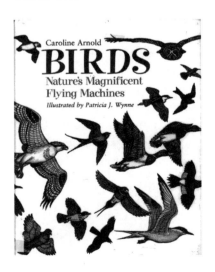

Birds: Nature's Magnificent Flying Machines by Carol Arnold (Watertown, MA: Charlesbridge, 2003); explanation, Flesch-Kincaid reading level 5.2.

How are birds able to fly? Colorful, detailed illustrations and clear explanations help students understand the adaptations and diversity of birds.

Field guides. Students will need access to appropriate field guides that clearly depict the birds' wings. For example:

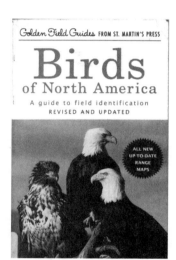

Birds of North America by Chandler S. Robbins, Bertel Bruun, and Herbert S. Zim (New York: St. Martin's Press, 2001); field guide, Flesch-Kincaid reading level 5.3.

Into the Air: An Illustrated Timeline of Flight by Ryan Ann Hunter (Washington, DC: National Geographic, 2003); reference, Flesch-Kincaid reading level 4.3.

Trace the history of flight back hundreds of million years ago to the very first flying creatures. Follow the development of flight all the way to today's aircraft. What will the future hold?

The Kids' Guide to Paper Airplanes by Christopher L. Harbo (Mankato, MN: Capstone Press, 2009); how-to, Flesch-Kincaid reading level 1.8.

Learn how to build 11 types of paper airplanes in this how-to book. Don't forget to study the folding techniques and terms on pages 6 and 7 before you begin!

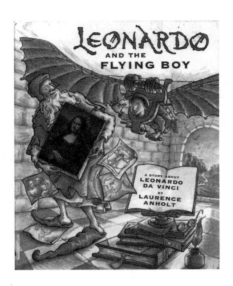

Leonardo and the Flying Boy by Laurence Anholt (Hauppage, NY: Barron's Educational Services, 2000); historical fiction, Flesch-Kincaid reading level 3.6.

Leonardo da Vinci is well-known for his paintings. But did you know that he was also a scientist and engineer whose ideas were ahead of his time? Learn about an invention that he worked on for many years—a flying machine!

MATERIALS

- Paper, cut in 6″ squares and 7″ × 10½″ rectangles (one each per pair of students)
- Computer with internet access (and access to YouTube)*
- Projector or interactive whiteboard
- Chart paper and marker (for whole-class work)
- Masking tape
- Metersticks
- Classroom library with flight-related books, including books about flying organisms (optional)
- Document camera (optional)
- Video camera (optional)
- Copies of supporting documents

Note: If access to YouTube is blocked, download the videos on a computer at home and save the files to a USB flash drive. The videos will be accessible without connecting to the YouTube site directly.

SUPPORTING DOCUMENTS

- Paper airplane data sheet
- DVTA graphic organizer

SAFETY CONSIDERATIONS

Instruct students to stand out of the path of the paper airplanes, and tell them not to launch the planes until all students are out of the way.

Come Fly With Me
Inquiry Unit

ENGAGE

In the engage phase, students are introduced to the topic of flight and are invited to compare flying organisms (birds) with man's early attempts at flight.

1. Read aloud the first two pages of *Leonardo and the Flying Boy* to introduce the idea that humans have always wanted to be able to fly.

    QR Code 1

2. Show the silent video clip "Early Flight" (*http://bit.ly/1athxT* [QR Code 1], 1:36). After viewing, discuss the clip with students, using guiding questions such as "Why didn't these attempts work?" and "Were any of the attempts more effective than others?" to help students begin to focus on the factors affecting flight.

3. Show the video clip "Birds in Flight Up Close and Personal" (*http://bit.ly/5qkOXo* [QR Code 2], 0:41). After viewing, discuss the clip with students, asking them, "How does the birds' flight compare to the attempts by people to fly?" and "How are birds able to fly?"

    QR Code 2

4. Read pages 3–11 of *Birds: Nature's Magnificent Flying Machines* and discuss adaptations that allow for flight: light bodies, hollow bones, wings, and feathers. Create a list of these adaptations on chart paper. As you read, take the time to have students complete the simple lift activity described on page 7.

5. Focus student attention on wings by asking, "How do birds' wings help them fly?" and "How does the shape of a bird's wing affect its flight?" Write these questions on chart paper and record student ideas underneath each.

Assess this phase: At this stage of the inquiry, the focus should be on formative assessment. Students should enter the next phase of the inquiry with a basic understanding that birds are adapted for flight in several ways (light bodies, hollow bones, wings, feathers). If students do not know or name all of these adaptations initially, they should be able to after the read-aloud and discussion around *Birds: Nature's Magnificent Flying Machines*.

EXPLORE

During the explore phase, students will test a factor that affects flight patterns and distance traveled of a paper airplane—the size of the wing. Students will construct and test two types of airplanes—one with long wings and one with short wings. They will record quantitative (distance

12

traveled) and qualitative (description) data about the flight path and the distance traveled. They will also sketch the flight path of each plane over multiple trials.

1. Explain to students that they will first investigate the relationship between wings and flight with paper airplanes, and then will use what they've learned to study the flight of various birds. Explain that they will test two types of planes—one with long wings and one with short wings.

2. Introduce the how-to book *The Kids' Guide to Paper Airplanes*, and explain that students will follow the directions to make two paper airplanes, the "Angry Finch" (pages 22–23) and the "Silent Huntress" (pages 24–25). Discuss with students the importance of controlling variables (such as the type of paper used and the shape of the plane) and how following a specific protocol for each plane does, in part, accomplish this. *Note:* At this point, students are not expected to connect the planes or their wing lengths to actual birds. We recommend downplaying the names if students inquire.

3. Provide time for students to construct the planes. Working in pairs ensures that all students have a chance to follow the procedures and build at least one plane. How you provide support for students during this period depends on your classroom situation. If you have a document camera, you might project the directions for the whole class to see at once. If you have access to multiple copies of the text, you might allow students to consult the books and work at their own pace. If you only have access to one copy, you might provide oral directions to the whole class or work with a small group of students at a time. If students work in small groups, designate group leaders who are comfortable folding the planes and who can make sure that the other members of their groups are able to successfully build the planes.

4. Take students to the gymnasium or hallway to practice flying the planes. Share the flying tips given for each plane in *The Kids' Guide to Paper Airplanes*, and discuss the importance of standardizing throws as much as possible (e.g., arm position, strength of throw, amount of follow-through). Students should work in pairs to fly the planes, attempting to develop a "standard" throw. Circulate among students to provide assistance to those having difficulty flying their planes. After a few trials, students will begin to get a sense for the flight pattern of each plane and will be able to judge if a trial is not representative of the plane's typical flight pattern. (The "Angry Finch," which has shorter wings, curves, swoops, and dives; the "Silent Huntress," which has longer wings, glides in a smoother, straighter manner and travels greater distances.)

5. Gather the class back together and explain that students will fly each plane five times, observing the path that the plane takes and the overall distance it travels. Each student will record quantitative (distance traveled) and qualitative (description of the flight path) data on the data collection sheet. Discuss how to measure the distance traveled with students. We recommend

using masking tape to mark the starting point and ending point and measuring between those points in a straight line. Again, circulate while students are conducting their trials, providing assistance and answering questions as needed. Keep students focused on gathering data by asking simple guiding questions:

- What did you notice about the way the plane flies?
- How would you describe that flight pattern?

You may also choose to use a video camera to record the planes in flight. This provides a permanent record of the flight paths of the two planes and can be used by students later on in the investigation.

6. When all pairs have completed their trials and drawings, gather the class together, and return to the classroom if you were working in the hallway or gymnasium. Explain that students will use their data to create an infographic depicting the flight patterns of the planes. Discuss what this drawing should look like, providing the following suggestions: create a key for each plane, using colors or dashed and solid lines to represent the paths of each plane; label each flight path with the trial number (1–5 for each plane); include points of reference from the gym or hallway in the sketch to convey a sense of distance; use straight, curved, and/or looped lines to represent the path the plane actually traveled. A sample infographic is shown in Figure 12.2.

FIGURE 12.2. SAMPLE INFOGRAPHIC DEPICTING FLIGHT PATTERNS

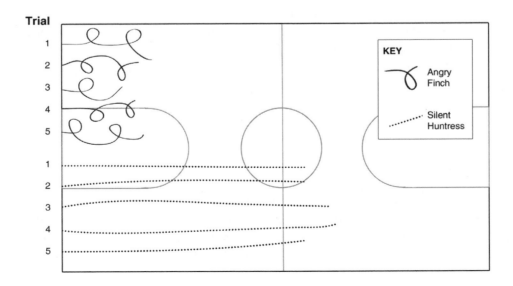

7. Allow students to share their sketches and findings in small groups before conducting a whole-class discussion. Did all their trials produce consistent results? If not, what other factors might have influenced the results?

8. Ask students to think-pair-share about what they learned from the investigation. How does the wing of an airplane influence its flight? Help students make evidence-based claims by posting a large T-chart, like the one illustrated in Figure 12.3, on chart paper. In the left-hand column, list evidence from the investigation. In the right-hand column, generate claims (conclusions) that are supported by the evidence. (Note: Think-pair-share is a three-step process that begins with students independently thinking about the prompt. Students then pair up to discuss their thinking and identify the best response. The process concludes with each pair sharing their thinking with the whole group.)

FIGURE 12.3. SAMPLE T-CHART LISTING CLAIMS AND EVIDENCE

Evidence	Claim
The "Angry Finch" plane did not fly in a straight line.	Shorter wings are good for turning.

Leave this chart posted through the remainder of the inquiry. Students will need to refer back to it during the next few phases of the inquiry.

Assess this phase: Again, formative assessment should be used during this phase. Monitor students' activity to ensure that all are successful in building their planes, developing standard throws, and collecting quantitative and qualitative data. If students struggle with these activities, you may opt to conduct them as a whole-class activity, or work with a small group that is finding the activity particularly challenging. Also monitor students' work as they create the infographics. Provide suggestions and support to students to help them clearly represent their data in visual form. Finally, pay close attention to students' ability to generate claims (conclusions) from their evidence. If needed, use guiding questions such as "What can we conclude from this evidence?" to help them complete this activity.

EXPLAIN

In this phase, students will share what they've learned so far. Taking the time to organize and articulate their thinking will prepare them for the second explore phase, in which they apply what they've learned about flight to birds' wings.

Each student should already have created an infographic (a detailed drawing) showing each plane and its flight path. Students should now write a paragraph to accompany the infographic that explains what the student has learned about how wings affect the flight of a plane.

Assess this phase: The infographic and paragraph should be considered a formative assessment task. If students are not drawing appropriate conclusions about the relationship between wing shape and flight pattern, repeat the plane investigation or conduct new trials as a class.

EXPLORE

In the second explore phase, students investigate the relationship between birds' wings and their flight patterns. Students will use the knowledge gained from the paper airplane experience to make predictions about the flight of several birds. They will observe birds in flight via video clips and will revise their predictions based on this new evidence. Finally, they will confirm their observations with informational text.

1. Explain to students that they will now study birds and their wings. Examine the drawings of birds on pages 30–31 of *Birds: Nature's Magnificent Flying Machines*. How do the wings of the birds compare? Students may share a wide variety of observations. Accept all reasonable responses, but guide students toward recognizing differences in the shapes of the wings. Ask students, "Based on what you learned from the paper airplanes, how do you think this will affect the birds' flight?" You may wish to record student responses on chart paper and post them in a prominent place during this portion of the inquiry.

2. Introduce the DVTA and the associated graphic organizer. Introduce the first bird, a black grouse. Have students study the picture, called a *planform*, which depicts the bird as it would be seen from directly above. Ask students to describe the shape of the wings (shorter and rounded). Based on the findings from the airplane investigation, what prediction could they make about the grouse's flight? Students should think-pair-share and record predictions on their graphic organizers.   QR Code 3

3. Play the video of the grouse in flight (*http://bit.ly/sqsTd1* [QR Code 3]). Ask students the following guiding questions to help them make detailed observations:

- How was the bird flying?
- Was it flying fast or slow?
- Was it flying high in the sky or close to the ground?
- Was it doing anything while flying, like catching prey?
- Can you find a way to draw or describe its motion?
- What did it do with its wings?

Students should record their observations in the "Video Evidence" column of the graphic organizer. They should then revise their prediction based on their evidence.

4. Repeat this process for the remaining birds: albatross (*http:/bit.ly/98bLJ* [QR Code 4]); seagull (*http://bit.ly/vHvP1h* [QR Code 5]); and pheasant (*http://bit.ly/ruxUbX* [QR Code 6]). You may choose to add additional birds if you feel that students need more evidence to draw conclusions. Here are some suggestions:

   » Artic tern: *http://bit.ly/rHKvOT* [QR Code 7]
   » Eagle: *http://bit.ly/rsqwM2* [QR Code 8]
   » Turkey vulture: *http://bit.ly/vv41Q4* [QR Code 9]
   » Peregrine falcon and Gos hawk: *http://bit.ly/11iSHg* [QR Code 10]

QR Code 4 QR Code 5 QR Code 6 QR Code 7 QR Code 8 QR Code 9 QR Code 10

- Many other videos of birds are available at *www.arkive.org* [QR Code 11].

5. Ask students to think-pair-share about the following questions:

- Which birds flew the same the way?
- Do you notice any similarities in their wings?

QR Code 11

Record student ideas on chart paper. If students struggle to answer these questions, repeat the DVTA with additional examples (see the suggestions in item 4 above).

6. Confirm student findings by reading pages 12–21 of *Birds: Nature's Magnificent Flying Machines*. As you read, highlight the flight-related vocabulary used in the text (*gliding, soaring, hovering, darting*). Have students act out each word, and add the words and definitions to your classroom word wall. If you don't have a word wall, create a temporary one using a piece of chart paper.

7. Return to the graphic organizer, and invite students to generate sentences about the birds using the vocabulary words as well as the evidence from the video. To assist students in linking these parts together, consider providing a sentence stem such as *The (name of bird) (flying verb) because its wings are (shape of wing)*. For example, *The albatross soars because its wings are long and wide*. Add these sentences to the words and definitions on the word wall.

Assess this phase: Formative assessment is particularly important at this stage of the inquiry. Students must be able to interpret the evidence from the video, confirm that evidence by listening to the text, and demonstrate understanding by using appropriate vocabulary in sentences. Monitor students as they complete the DVTA to make sure they are focusing on relevant information and drawing appropriate conclusions based on the video evidence. If not, consider watching the videos a second time, and increase the number of guiding questions you ask to help focus student attention. If students have difficulty connecting the videos to the read-aloud, consider replaying the videos as your read each type of flight. Watching the videos a second time may also help students who have difficulty composing sentences with the flight-related vocabulary.

EXPLAIN

In the second explain phase, students will synthesize what they've learned about wing shape and flight.

1. Students should select a bird not studied in class from a field guide. Ask them to produce a product that demonstrates their understanding of how the wings affect the way in which it flies. The actual assignment can take a variety of different forms, including a paragraph and illustration, a flip book showing the bird in flight with an accompanying explanation, or a "slowmation" (*http://slowmation.com* [QR Code 12]) showing the bird in flight with an QR Code 12 accompanying explanation. Students should have access to the videos, materials, and charts from the earlier experiences. A reference library with books about flying organisms would also be helpful. You may also wish to allow students to search online for video clips of their birds in flight. The ARKive website (*www.arkive.org* [QR Code 11]) contains images, videos, and background information for thousands of plant and animal species found around the QR Code 11 world. Allow students time to share their products with others, either in small groups or as a whole class.

Assess this phase: The product created in this phase serves as summative assessment for the inquiry unit. Students should correctly describe the shape of the selected birds' wings and explain how the shape of the wings influences the way in which the birds fly. The "Science and Literacy Rubric" in Appendix 2 (p. 276) can be used to assess student performance, and the "Achievement Grading Standards" (also in Appendix 2, p. 277) can translate this into a numerical grade. If students fall below 75%, return to the explore phase for additional work before returning to the writing prompt.

EXPAND

In this phase, students will connect what they've learned about structure and function in living organisms (birds) to designed inventions.

1. Read *Into the Air* aloud and discuss how designed inventions like balloons, planes, and gliders mimic the structure and function seen in the natural world.

2. Students should each choose a designed invention that mimics the structure and function of the bird they studied in the explain phase. Have students research that invention and draw/ write how its design is similar to the birds they studied. A T-chart may help students organize their thinking while conducting research. Allow students time to share their research, either in small groups or with the whole class.

3. Conclude the unit by reading aloud the remainder of *Leonardo and the Flying Boy*.

Assess this phase: Student work in this phase also serves as summative assessment. Students should make connections between the structure and function of the invention in the same way they did for the bird studied in the explain phase. The "Science and Literacy Rubric" in Appendix 2 can be used to assess student performance, and the "Achievement Grading Standards" (also in Appendix 2) can translate this into a numerical grade. If students fall below 75%, use guiding questions to help them relate what they've learned about the structure and function of wings to designed inventions.

REFERENCES

Cardak, O. 2009. Science students' misconceptions about birds. *Scientific Research and Essays* 4 (12): 1518–1522.

Deinlein, M. n.d. Have wings, will travel: Avian adaptations to migration. Smithsonian National Zoological Park, Migratory Bird Center. *http://nationalzoo.si.edu/scbi/migratorybirds/fact_sheets/default.cfm?fxsht=4*

Ehrlich, P. R., D. S. Dobkin, and D. Wheye. 1988. Adaptations for flight. *www.stanford.edu/group/stanfordbirds/ text/essays/Adaptations.html*

Kazilek, C. J. 2009. Ask a biologist: Feather biology. Tempe, AZ: Arizona State University, School of Life Sciences. *http://askabiologist.asu.edu/how-do-birds-fly*

London, K. 2011. London zoo: Wing shape affects flight. *http://azdailysun.com/lifestyles/pets/article_0c4605ae-a31f-5743-adf1-4e61600464cf.html*

Stauffer, R. G. 1975. *Directing the reading-thinking process*. New York: HarperCollins.

Name_____ Date_____

PAPER AIRPLANE DATA

Airplane 1: Angry Finch
Wing Length: Short   Long   (circle one)

Trial Number	Distance Traveled	Description of Flight Path
1		
2		
3		
4		
5		

Name_____ Date_____

Airplane 2: Silent Huntress
Wing Length: Short   Long (circle one)

Trial Number	Distance Traveled	Description of Flight Path
1		
2		
3		
4		
5		

Name_____ Date_____

DVTA GRAPHIC ORGANIZER

Bird Name	Picture (Planform)	Prediction About Bird's Flight	Video Evidence	Revised Prediction Based on Video
Black grouse				
Albatross				
Seagull				
Pheasant				

Chapter 13
Drip Drop Detectives: Exposing the Water Cycle

OVERVIEW

Throughout the inquiry, students play the role of scientists who are investigating the question, *How can a drop of water travel around the world*? Students examine the water cycle: evaporation, condensation, infiltration, and transpiration. Then students develop a water cycle diagram based on what they have learned. Along with learning about the water cycle, students also learn about the importance of the reproducibility of results.

In a natural confluence of science and language arts, the scientists use their knowledge to write a newsletter in which they "expose" how a mere drop of water can travel around the world.

In Chapter 8, "Minds-on Matter: Phase Changes and Physical Properties," students observe and measure the physical properties of solids, liquids, and gases. They also measure the changes in temperature that occur during melting, freezing, and boiling. In addition, students are introduced to the study of ice cores, after which they engage in an engineering challenge to design a container that keeps an ice cube from melting. While the concept of phase changes is addressed in both Chapter 8 and this chapter, there are distinct differences in the approaches to the topic. This chapter deals strictly with water; Chapter 8 involves a variety of substances. These two units were designed to be independent of one another to ensure flexibility within any school's curriculum. Yet if your curriculum is designed to address the topics of the water cycle, phase change, and physical properties in the same grade level, you may want to consider working with your students to draw relationships between the two sets of investigations.

OBJECTIVES

- Investigate water cycle processes
- Construct a water cycle diagram from information found in nonfiction texts
- Write an article explaining the water cycle
- Recognize that the Earth's supply of freshwater is limited
- Gather and analyze water usage data

STANDARDS ALIGNMENT

National Science Education Standards
SCIENCE AS INQUIRY
- K–4, 5–8 Abilities Necessary to Do Scientific Inquiry

EARTH AND SPACE SCIENCE
- 5–8 Structure of the Earth System

SCIENCE IN PERSONAL AND SOCIAL PERSPECTIVES
- K–4 Types of Resources

Common Core State Standards for English Language Arts
INFORMATIONAL TEXT
- Grades 3–5 Key Ideas and Details
- Grades 3–5 Craft and Structure
- Grades 3–5 Integration of Knowledge and Ideas

WRITING
- Grades 3–5 Text Types and Purposes
- Grades 3–5 Research to Build and Present Knowledge

For a detailed standards alignment, see Appendix 3 (p. 282).

TIME FRAME
- Seven 45-minute class periods

SCIENTIFIC BACKGROUND INFORMATION

The water cycle, the primary focus of this inquiry, is a naturally occurring cycle in which water circulates through the hydrosphere, atmosphere, biosphere, and lithosphere. Water occurs on Earth as a solid, liquid, and vapor. The physical changes that occur as water passes through these phases are an important part of the water cycle. Water enters the atmosphere when it evaporates from surface waters such as lakes, rivers, and the ocean. The water vapor then cools and condenses to form clouds. Precipitation follows, returning the water to the surface of the Earth. On Earth's surface the water may infiltrate into groundwater reservoirs, run off into surface waters, or be taken in by plants. Plants then release the water back into the atmosphere through transpiration.

The Sun's energy drives the water cycle. Radiant energy from the Sun warms the oceans, lakes, and other bodies of water on Earth. When the warm water on the surface of these bodies of water has sufficient energy, it evaporates. Solar energy also increases the rate at which plants release water to the atmosphere through transpiration. Energy from the Sun also plays a role in the formation of convection cells that produce wind. As warm air in the atmosphere (including water vapor) rises, cool air takes its place. Wind plays an important role in moving water vapor in the atmosphere.

Reproducibility of results is a critical aspect of science. Data that are reproducible are considered less tentative than data that cannot be reproduced. Data that are not reproducible are often discarded. Reproducing the original results provides additional evidence to support the claims. Inability to reproduce results casts doubt on the original findings and associated claims.

From the perspective of elementary science education it is appropriate to discuss the idea of reproducibility of results or independent confirmation using terms children may be more familiar with. Terms such as *verify*, *confirm*, or *double-check* are acceptable synonyms for reproducibility of results.

MISCONCEPTIONS

Surprisingly little research has been done on children's conceptions about the water cycle. Shepardson and colleagues (2009) found that children's conceptions of the water cycle can be organized into four primary ideas. These include

- The water cycle as a mechanism for water storage, transformation, and transportation of water
- The water cycle as a mechanism for water storage and transformation
- The water cycle as a weather event
- The water cycle as an entity, a water source

Students holding the first view had a fairly sophisticated and scientifically accurate view of the water cycle. Students holding the remaining views had misconceptions about the water cycle. Table 13.1 highlights some of these misconceptions.

TABLE 13.1. COMMON MISCONCEPTIONS ABOUT THE WATER CYCLE

Common Misconception	Scientifically Accurate Concept
The water cycle only involves precipitation, evaporation, and condensation.	The water cycle involves a variety of processes including evaporation, condensation, infiltration, precipitation, transpiration, runoff, and sublimation.
The water cycle only includes rain and other forms of precipitation.	The water cycle involves the storage, transformation, and transportation of water.
The water cycle involves rivers, lakes, and other waters on the surface of the Earth.	The water cycle includes all surface waters, frozen water (sea ice, glaciers, ice sheets, icebergs, permafrost), runoff, and groundwater.

Shepardson and colleagues also examined fourth- and fifth-grade science textbooks and found that the water cycle diagrams in these books typically included landscapes that featured oceans and mountains. Limiting diagrams to a single landscape may make it difficult for children to generalize the water cycle to a global view. Exposing students to multiple representations of the water cycle and to the range of processes involved may help address some of these misconceptions.

TEXT SET

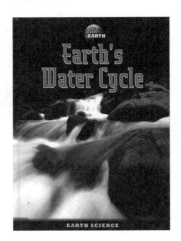

Earth's Water Cycle by Amy Bauman (Pleasantville, NY: Gareth Stevens, 2008); reference, Flesch-Kincaid reading level 5.5.

Each page is packed with information and colorful photographs and sometimes a drawing. As students read about the water cycle, they encounter the expected—such as liquids, gases, and solids—and often the unexpected—such as a diagram of a water molecule and photos of blood plasma and watermelon.

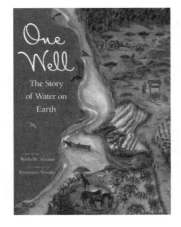

One Well: The Story of Water on Earth by Rochelle Strauss (Tonawanda, NY: Kids Can Press, 2007); explanation, Flesch-Kincaid reading level 6.0.

Water on Earth is precious and must be treated as such. This is a message students will understand as they read statistics and descriptions of water use and availability around the planet. The vividly colored illustrations are charming.

DRIP DROP DETECTIVES:
EXPOSING THE WATER CYCLE

The Life and Times of a Drop of Water by Angela Royston (Chicago: Raintree, 2006); reference, Flesch-Kincaid reading level 3.7.

A drop of water makes its way through rain, puddles, clouds, and so on; in other words, it makes its way through the water cycle. The description of the water cycle goes beyond the basics to include water passing through animals, traveling through a water treatment plant, and more.

The Snowflake: A Water Cycle Story by Neil Waldman (Brookfield, CT: Millbrook, 2003); narrative expository, Flesch-Kincaid reading level 6.0.

Readers follow the journey of a snowflake, month by month, and watch it change from one form of precipitation to another through the seasons and over great distances. The writing is graceful, and the watercolor illustrations are perfectly suited to the text.

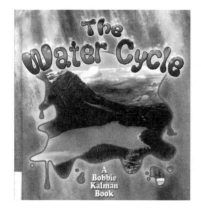

The Water Cycle by Bobbie Kalman and Rebecca Sjonger (New York: Crabtree, 2006); reference, Flesch-Kincaid reading level 4.2.

Evaporation, condensation, precipitation—this book explores it all. It describes freshwater, salt water, and even polluted water. Engaging photos picture a range of subjects, including a rainbow, a ring of skydivers above a cloud, and brightly colored fish in the ocean.

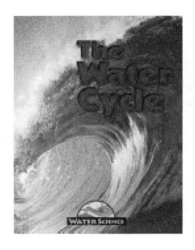

The Water Cycle by Frances Purslow (New York: AV² by Weigl, 2001); reference, Flesch-Kincaid reading level 5.1.

This representation of the water cycle includes forms of condensation and precipitation.

The Water Cycle by Marcia Zappa (Edina, MN: ABDO, 2011); reference, Flesch-Kincaid reading level 3.9.

For the most part, left-hand pages present information about water and the water cycle, and right-hand pages feature neat photographs with boxed captions. The sentences are short, and the font is large. The diagram of the water cycle is simple and easy to follow.

The Water Cycle by Rebecca Olien (Mankato, MN: Capstone, 2005); reference, Flesch-Kincaid reading level 2.9.

Like every other book in this text set, this book does an excellent job of explaining the basics of the water cycle. And like the other books, the illustrations are appropriate and interesting. The reading level is lower than that of the other books, which makes it a good choice for less proficient readers. But even though the reading level is lower, the writing is smooth and natural sounding.

MATERIALS

- Graduated cylinders or measuring cups, enough for two teams
- Water
- Ice
- Two plants
- Desk lamp with incandescent bulb
- Six small plastic containers with lids
- Three small shallow plastic containers without lids
- Styrofoam plates
- Three 2 L bottles with the bottoms removed or large transparent plastic bags to completely cover plants
- Four 1 L bottles with the tops removed so they can be inverted and used as funnels and small holes drilled in the lids
- Plastic wrap
- Markers
- Soil, approximately 1 c.
- Sand, approximately 1 c.
- Gravel, approximately 1 c.
- Chalk
- Computer with internet access
- Digital camera (optional)
- Copies of supporting documents

SUPPORTING DOCUMENTS

- "Informant Investigation Sheet" for teams 1–4
- Tip cards
- "Checking What We Know" worksheet
- "INTK Article Map"
- "Anticipation Guide: One Well: The Story of Water on Earth"

SAFETY CONSIDERATIONS

Students should be careful while working with the incandescent lightbulb.

Drip Drop Detectives Inquiry Unit

ENGAGE

In the engage phase students are challenged to test several water cycle processes that have been described by confidential informants.

1. Begin the inquiry by setting the stage for the roles you and your students will play throughout the inquiry. Tell the students that you have just come across an interesting story about a drop of water that has reportedly traveled around the world. Share that you are having a hard time believing that a drop of water can travel around the world and you have decided that as a class they should look into this incredible story. Inform the students that you will now share the story with them.

2. Read aloud *The Snowflake* but stop after the December page; do not read the last two pages. While you are reading, follow the progress of the snowflake/water drop by placing stickers on a world map showing where the water has been. Interject comments along the way that illustrate your skepticism that a drop of water can travel around the world. You might say something like "I just find it hard to believe a drop of water could travel from a lake in Wyoming to a city in Colorado. How would a drop of water travel? Water cannot fly, it has no wheels, water can't move!" Table 13.2 (p. 199) offers a set of locations you could use. Place a sticker on the map while reading the corresponding page of the book. Remember to insert comments that help draw the students in and instill in them a need to find out how a drop of water can travel around the world. Invite the class to share their questions and comments as you read.

3. After reading the story, share that four confidential informants have provided tips about how the droplet traveled around the world. Each informant also provided information about tests that can be performed to see if the tip is accurate.

4. Explain to the students that they will be acting as scientists investigating the question, *How can a drop of water travel around the world?* They will then publish a newsletter called *Important News to Know,* or INTK. As scientists they will work in teams to investigate the tips provided by the confidential informants. Teams will follow leads/tips provided by the confidential informants. Each team will investigate a different tip.

    • Tip 1: The water is moving from place to place by changing from a liquid to a gas.

    • Tip 2: The water is moving from place to place through underground reservoirs.

    • Tip 3: The water is moving from place to place by changing from a gas to a liquid.

    • Tip 4: The water is moving from place to place through plants.

DRIP DROP DETECTIVES:
EXPOSING THE WATER CYCLE

13

TABLE 13.2. SUGGESTED LOCATIONS FOR SNOWFLAKE TRACKING

Location	Month	Comment
Teton Range of Rocky Mountains	January	Okay, I can see that the water droplet started its journey as a snowflake in the Rockies.
Yellowstone Lake, Wyoming	February	The droplet is now in Yellowstone Lake in Wyoming. How can that be?
South Platte River, Denver, Colorado	April	I really don't get this. How did this droplet get from a lake in Wyoming to a river in Denver, Colorado?
Platte River, central Nebraska	May	Now I've heard that they grow cabbage in Nebraska. That I understand, but I'm having some trouble seeing how the water droplet ended up there.
Seine River, France	July	This water droplet is in Europe! I've never been to Europe and this droplet makes it over the Atlantic! Unheard of!
Paris, France	September	Do you think that if I could turn into a droplet of water I could make it to Paris, France?
Pacific Ocean	October	How did the droplet get all the way to the Pacific Ocean?
Hawaii	November	Hawaii! Another of Earth's most desirable vacation destinations. I have to say the droplet ends up in some nice places. But how?
Mountain range in western United States	December	Look at that. According to this account, a droplet of water has traveled around the world. I've got to find out if this is possible. And I need your help.

5. After completing their investigations, each team will post their results on the "fact board" using the format shown in Table 13.3 (p. 200). After sharing their results, teams will confirm the findings of their colleagues. The newsletter's special edition on water will be constructed using confirmed facts on the fact board.

Assess this phase: This phase serves to introduce students to the idea of water traveling around the world. Student comments and reactions to the text and your comments serve as an informal formative assessment that will provide insight into their current knowledge of phase changes and the water cycle. Students may reveal misconceptions as you are reading. It is not necessary to address misconceptions at this time. As the unit progresses, opportunities may arise to address the misconceptions.

TABLE 13.3. FACT BOARD

Tip	Test 1 Evidence	Test 2 Evidence	Tip Confirmed?		Connections
			Test 1	Test 2	
Changing from a liquid to a gas					
Moving through underground reservoirs					
Changing from a gas to a liquid					
Moving through plants					

EXPLORE

In the explore phase students investigate evaporation, infiltration, condensation, and transpiration.

Advance preparation: Set up four stations, one for each of the tip card investigations. If the class is large, you may want to set up more than one station for each investigation. Alternatively, set the stations up in a work area large enough to accommodate two teams.

Prepare the fact board on chart paper or on a section of the chalkboard or whiteboard where it can be left for the duration of the inquiry. *Note:* Environmental factors such as humidity can affect the rate of evaporation. Prior to the investigation do some trial runs with different volumes of water. Measure several different volumes of water, from 5 ml to 30 ml. Pour each into a small shallow plastic container. Let them sit overnight. The next day measure the remaining water. Ideally a noticeable amount of water will have evaporated. Determine which volume will return the most noticeable results. Team 1 should use this volume in their investigation.

Day 1

1. Organize the students into four teams (more if the class is large). Give each team a tip card and direct them to the corresponding workstation. Each team should work through the inquiry as written on its Informant Investigation Sheet. Circulate through the room providing assistance, observing student progress, and asking the guiding questions listed for each team.

TEAM 1

Tip 1: The water is moving from place to place by changing from a liquid to a gas.

This team will use "Informant Investigation Sheet: Team 1" and the following guiding questions:

- What do you notice about the water in the cups?

- How do you explain the changes you have observed? (This question is for day 2.)

TEAM 2

Tip 2: The water is moving from place to place through underground reservoirs.

This team will use "Informant Investigation Sheet: Team 2" and the following guiding questions:

- What do you notice about the water in the bottles?

- How does the amount of water in each bottle compare?

- How do you explain the changes you have observed?

- The water is trapped in the bottle and therefore cannot move from place to place as the informant suggested. How is this different from what would happen in nature?

TEAM 3

Tip 3: The water is moving from place to place by changing from a gas to a liquid.

This team will use "Informant Investigation Sheet: Team 3" and the following guiding questions:

- What do you notice about the containers?

- How does the amount of water on the containers compare?

- How do you explain the changes you have observed?

Team 4 *Tip 4: The water is moving from place to place through plants.*

This team will use "Informant Investigation Sheet: Team 4" and the following guiding questions:

- What do you notice about the bottles?

- How does the amount of water in the bottles compare?

- How do you explain the changes you have observed? (This question is for day 2.)

Teams 2 and 3 will be able to complete their investigation in one session. Teams 1 and 4 will need to let their investigations sit overnight. They will collect their data the next day.

Day 2

1. Instruct Teams 1 and 4 to complete the investigations they started the day before. During this time Teams 2 and 3 can record their information on the fact board, leaving the confirmation and connections columns blank. After all teams have finished collecting their data and the fact board, bring the class back together for a class discussion.

2. Ask each team to share their findings. Each team will describe their procedure, share their data, and discuss if, based on the evidence they have, the tip is confirmed or rejected. Encourage students from other teams to ask clarifying questions. After their presentations each team will

indicate on the fact board whether their tip was confirmed or rejected.

3. When all teams have finished, call attention to the fact that the investigations of Teams 2, 3, and 4 included controls. Ask the class to recall what a control is and why it is important. Follow up by asking what the controls are for the investigations conducted by Team 1.

After the discussion, share that you are very pleased with their findings but you want to know more. You want the student scientists to confirm that the same thing will happen under different conditions; a different team will now conduct each investigation outside. *Note:* The temperature outside can greatly affect the results of the investigations. If it is too cold (e.g., near freezing), the water may freeze rather than evaporate. We suggest conducting this inquiry unit in the spring or fall when temperatures are warmer.

4. Teams will now confirm the findings of their fellow scientists by performing similar tests outside. Students have more freedom to design their investigations in this step. Their investigations should build on the ones previously conducted. Possible investigations are provided here for guidance.

TIP 1: EVAPORATION
Materials: graduated cylinder or measuring cups, water, chalk, digital camera (optional)

The team will create puddles outside on pavement or a sidewalk. They will create two puddles, each with the same volume of water. One should be in the shade, the other in the full sun. Carefully pour the water onto the pavement. Draw a chalk line around the puddle. Repeat the process for the second location. Wait 15 minutes, and then draw a second chalk line around the puddle. Repeat the process one more time. Document the changes in the puddle with a diagram or digital photography.

TIP 2: CONDENSATION
Materials: six small plastic containers with lids, ice, water, Styrofoam plates, markers

Select two test sites, one in the shade and one in the full sun. Prepare the test materials. Label two containers "Ice," another two "Water," and the final two "Control." Fill the ice containers half full with ice. Quickly cover the containers and place each on a separate Styrofoam plate. Observe the outside of the container and record if it is damp or dry. Fill the water containers half full of water. Cover the containers and place each on a separate Styrofoam plate. Observe the outside of the container and record if it is damp or dry. Place the lids on the control containers and place each on a separate Styrofoam plate. Observe the outside of the container and record if it is damp or dry. Place one set of systems in the sun and one set in the shade. Allow the systems to sit for 30 minutes. After 30 minutes have passed, observe the outside of the containers and record if each is damp or dry.

TIP 3: TRANSPIRATION

Materials: two plants, three 2 L bottles with bottoms removed

Label two bottles "Plant" and two bottles "Control." Set one plant in a sunny location where it will not be disturbed. Cover it with one of the bottles labeled "Plant." Set one of the bottles labeled "Control" next to it. Set another plant in a shady location where it will not be disturbed. Set the other bottle labeled "Control" next to it. Cover this plant with the other bottle labeled "Plant." Let the bottles and plants sit over night. Observe and record data the following day.

TIP 4: INFILTRATION

Materials: graduated cylinder or measuring cups, water, digital camera (optional)

Select three different sites for testing. Potential sites include grassy lawns, gravel lanes, sandboxes or sandy soil, pavement, or sidewalks. Pour equal volumes of water on each surface. Observe and document what happens.

5. Each team should record their results on the fact board after they have finished collecting data.

6. After all teams have recorded their results on the fact board, bring the class together to discuss the findings.

7. After discussing the results, refer back to *The Snowflake* and the map. You can either read the entire book again or read only the pages that allow students to make specific connections between the story and the four confidential informants' tips. As you are reading, pause to record the month and movement of the water drop in the "Connections" column of the fact board. Record this information in the row that corresponds with the process that occurred. For example, in March the droplet infiltrates into an underground stream, so write "March" and the movement in the "Moving through underground reservoirs" row in the "Connections" column (see Table 13.4). The idea is to connect what the students learned in their investigation back to the journey the water made around the world.

8. Each team should discuss the findings from all of the investigations and develop an explanation for how a drop of water can travel around the world. Teams should share the explanation with the teacher before verifying their thinking with the texts. Students should use the "Checking What We Know" worksheet and reference books from the text set (*Earth's Water Cycle, The Life and Times of a Drop of Water,* and the four books entitled *The Water Cycle*) to refine their thinking and discover the appropriate terms for the processes they investigated. The texts will also bring in the process of precipitation. As the teams are working, circulate through the room asking these guiding questions:

   • What happens to the water in each of the water cycle processes?

TABLE 13.4. COMPLETING THE CONNECTIONS COLUMN ON THE FACT BOARD

Tip	Test 1 Evidence	Test 2 Evidence	Tip Confirmed?		Connections
			Test 1	Test 2	
Changing from a liquid to a gas					
Moving through underground reservoirs	Water moves through the sand, soil, and gravel into the bottle below.	Water goes into the ground from the lawn and gravel parking lot. It stays on top of the sidewalk.	Yes	Yes	March, water went into an underground stream
Changing from a gas to a liquid					
Moving through plants					

- Which of the water cycle processes involves water vapor?
- How is energy from the Sun involved in the water cycle?

Offer assistance where needed. Be sure to point out the water cycle diagrams that appear in most of the books. This will provide helpful examples for the students when they are developing their own diagrams.

9. Each of the investigations represents a model of one part of the water cycle. Ask the class to recall what a scientific model is and what purpose models serve. Then ask how each of the investigations is similar to and different from what actually happens in nature.

Assess this phase: As students build an understanding of the water cycle in this phase, formative assessment is critical. Monitor students' completion of the investigations and in particular, their responses to the guiding question, "How do you explain the changes you have observed?" These will provide insight into their understanding of the water cycle–related phenomena (evaporation, infiltration, condensation, and transpiration) being modeled by the students. If students have an incomplete or incorrect understanding of the phenomena, do not correct them at this point, but pose additional questions to encourage them to consider the evidence more deeply. Students will confirm their explanations with the texts later on in the explore phase, ensuring that they develop correct understandings of the phenomena and the water cycle as a whole.

EXPLAIN

In the explain phase students share their understanding of the water cycle through the development of a water cycle diagram and an article that answers the question, *How can a drop of water travel around the world?*

1. Using information from the fact board, the team's explanation, and the texts, each team will develop a water cycle diagram that illustrates evaporation, condensation, infiltration, transpiration, and precipitation. Circulate around the room asking teams the guiding questions below. Remind the students about the water cycle diagrams they observed in the books they used to check their knowledge.

    *   How can you help readers understand the water cycle?
    *   How can you show the water cycle as a diagram?
    *   How can you organize your thinking?
    *   How might you make this more clear?
    *   How can you show that the water cycle is ongoing?
    *   What information should be included and what can be left out?

2. Each student will then write an article for the special edition of *Important News to Know,* in which they answer the question, *How can a drop of water travel around the world?* Students will use the "INTK Article Map" to help organize their thoughts.

Assess this phase: The water cycle diagram and article serve as summative assessment for the inquiry. Student work should reflect an understanding of the water cycle as well as the phenomena (evaporation, transpiration, infiltration, and condensation) that make up this cycle. The "Science and Literacy Rubric" in Appendix 2 (p. 276) can be used to assess student performance, and the "Achievement Grading Standards" (also in Appendix 2, p. 277) can translate this into a numerical grade. If students fall below 75%, return to the explore phase for additional work before returning to the writing prompt.

EXPAND

In the expand phase students begin to consider water as a natural resource that should be conserved. They examine their own use of water and identify steps to conserve water.

1. Have students complete the anticipation guide for this text before you begin reading, and tell the students that they will be learning about how much water there is on Earth and how we depend on water for our survival. Read aloud the following pages from *One Well: The Story of Water on Earth:* 4, 7, 8, 16, 19, 20, 21, 23, and 26. While you are reading, students should listen for evidence that supports or rejects their prediction. After you have finished reading, students will revise their predictions based on evidence from the text. Discuss the reading with the class

asking them what surprised them, what they expected, and what they would like to know more about. Follow this by telling students they will be monitoring their water use for the next few days.

2. Have students keep a simple water diary for several days. In their diary they should record when and how they used water. Discuss with students uses of water they will record, such as drinking, brushing their teeth, bathing, flushing the toilet, washing the dog, and watering plants. Students will determine how to record the information in their diaries.

3. Students will share their data after the preset time for data collection has passed. Begin by asking students to share how they used water. List the ways they used water on chart paper or the chalk/whiteboard. After the list has been compiled, create a frequency table by having students come up individually and put tally marks next to the ways they used water, with one tally mark for each use. If they did the dishes three times they should put three tally marks next to "Washed the dishes." Now total the tally marks for each different use of water. Discuss the class data using the following questions as a guide.

   • What surprises you about the data?

   • What does not surprise you?

   • This data represents (number) days. How much water might we use in a month? A year?

   • How can we reduce our water use?

4. Brainstorm ways to reduce water use at home and in school. Following the discussion students will work in teams to conduct an internet search for additional water conservation tips. Have each student make a list of five actions they can take to reduce water use.

Assess this phase: Formative assessment is used in this phase. Anticipation guides, water diaries, and lists of water conservation actions all provide evidence of student engagement with the topic and understanding of water use and conservation.

REFERENCE

Shepardson, D. P., B. Wee, M. Priddy, L. Schellenberger, and J. Harbor. 2009. Water transformation and storage in the mountains and at the coast: Midwest students' disconnected conceptions of the hydrologic cycle. *International Journal of Science Education* 31 (11): 1447–1471.

Name_____ Date_____

INFORMANT INVESTIGATION SHEET: TEAM 1

Use these supplies and directions to investigate your tip: The water is moving from place to place by changing from a liquid to a gas.

Supplies

- Three small, shallow plastic containers
- Water
- Graduated cylinder or measuring spoons
- Plastic wrap
- Marker
- Lamp with incandescent lightbulb

Directions

1. Label one container "Covered," another "Light Energy," and the other "Uncovered."
2. Pour _____ ml or _____ tbsp. of water into each plastic container. Record the beginning volume.
3. Tightly cover the container labeled "Covered." Place it in a location where it will not be disturbed.
4. Place the container labeled "Light Energy" under a desk light with an incandescent lightbulb in a location where it will not be disturbed.
5. Set the container labeled "Uncovered" in a location other than a windowsill where it will not be disturbed.
6. Let the containers sit overnight.
7. Retrieve the containers the next day and carefully measure the remaining water. Record the remaining volume of water.
8. Prepare your team presentation. The presentation must include your tip, a description of your investigation, your findings, and your conclusion.

Name_____ Date_____

	Beginning Volume	Ending Volume	Change
Covered			
Uncovered			
Light Energy			

Name_____ Date_____

INFORMANT INVESTIGATION SHEET: TEAM 2

Use these supplies and directions to investigate your tip: The water is moving from place to place through underground reservoirs.

Supplies

- Four 1 L bottles with the tops removed and small holes drilled in the lids
- Soil
- Sand
- Gravel
- Graduated cylinder
- Marker

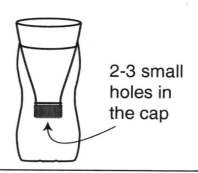

2-3 small holes in the cap

Directions

1. Label the side of the bottom half of one bottle "Soil," another "Sand," another "Gravel," and the other "Control."
2. Invert the tops of the bottles and place them over the bottoms of the bottle.
3. Fill the inverted tops so that they are full to the top with the corresponding earth material.
4. Slowly pour 100 ml of water over each earth material.
5. After five minutes take the top off the bottle and set it aside.
6. Measure and record the water that has collected in the bottom of the bottle.
7. Calculate the change in volume.
8. Prepare your team presentation. The presentation must include your tip, a description of your investigation, your findings, and your conclusion.

	Volume of Water Poured Into Bottle	Volume of Water in Bottom of Bottle	Change in Volume
Control			
Soil			
Sand			
Gravel			

Name_____ Date_____

INFORMANT INVESTIGATION SHEET: TEAM 3

Use these supplies and directions to investigate your tip: The water is moving from place to place by changing from a gas to a liquid.

Supplies

- Three plastic containers with lids
- Ice
- Water
- Styrofoam plates
- Marker

Measurement Scale

Number	Amount of Water
0	No water
1	Very little water
2	Small water droplets
3	Large water droplets
4	Water droplets running down the side of the container
5	Large or small puddles of water on the plate

Directions

1. Label one container "Ice," another "Water," and the other "Control."
2. Fill the container labeled "Water" half full of water. Cover the container and place it on one of the Styrofoam plates.
3. Place the lid on the container labeled "Control" and place it on one of the Styrofoam plates.
4. Fill the "Ice" container half full with ice. Quickly cover the container and place it on one of the Styrofoam plates.
5. Notice if there are any water droplets on the outside of the containers. Record your observations. Use the measurement scale at the top of the page as a guide.
6. Place the three containers on their Styrofoam plates in a location where they will not be disturbed.
7. Allow the containers to sit for 30 minutes.
8. After 30 minutes have passed, collect the containers and observe if any additional water has collected on the outside of the container. Record your observations.
9. Prepare your team presentation. The presentation must include your tip, a description of your investigation, your findings, and your conclusion.

Name_____ Date_____

	Water on the Container Before	Water on the Container After	Change (increase, decrease, no change)
Water			
Ice Water			
Control			

Name_____ Date_____

INFORMANT INVESTIGATION SHEET: TEAM 4

Use these supplies and directions to investigate your tip: The water is moving from place to place through plants.

Supplies

- Two plants
- Three 2 L bottles with bottoms removed
- Desk lamp with incandescent bulb
- Marker

Measurement Scale

Number	Amount of Water
0	No water
1	Very little water
2	Small water droplets
3	Large water droplets
4	Water droplets running down the inside of the container
5	Large or small puddles of water inside the container

Directions

1. Label one bottle "Plant," another "Light Energy," and the other "Control."
2. Set one plant in a location where it will not be disturbed. Cover it with the bottle labeled "Plant."
3. Set the bottle labeled "Control" next to it.
4. Set another plant and the desk lamp in a location where it will not be disturbed. Cover the plant with the bottle labeled "Light Energy." Position the light above the plant.
5. Notice if there are any water droplets on the inside of the bottles. Record your observations. Use the measurement scale at the top of the page as a guide.
6. Let the bottles and plants sit overnight.
7. The next day, observe if any additional water has collected on the inside of the bottles. Record your observations.
8. Prepare your team presentation. The presentation must include your tip, a description of your investigation, your findings, and your conclusion.

Name_____ Date_____

	Water on Inside of Bottle Before	Water on Inside of Bottle After	Change (increase, decrease, no change)
Control			
Plant			
Light Energy			

Name_____ Date_____

TIP CARDS

Cut apart and distribute one card to each group.

Tip 1

The water is moving from place to place by changing from a liquid to a gas.

Tip 2

The water is moving from place to place through underground reservoirs.

Tip 3

The water is moving from place to place by changing from a gas to a liquid.

Tip 4

The water is moving from place to place through plants.

DRIP DROP DETECTIVES:
EXPOSING THE WATER CYCLE

Name_____ Date_____

CHECKING WHAT WE KNOW

Team's Explanation

_____

_____

_____

_____

Evidence from the book that confirms our explanation:
Things we could not confirm:
New information:

Name_____ Date_____

INTK ARTICLE MAP

How can a drop of water travel around the world?

Topic Sentence: _____

_____

Detail: Evaporation
_____
_____
_____

Detail: Condensation
_____
_____
_____

Detail: Precipitation
_____
_____
_____

Detail: Infiltration
_____
_____
_____

Detail: Transpiration
_____
_____
_____

Conclusion:_____

Name_____ Date_____

ANTICIPATION GUIDE: *ONE WELL: THE STORY OF WATER ON EARTH*

Before the book is read, check if you agree or disagree with each statement. Use what you have learned about the water cycle to help you with your predictions. While the book is being read listen closely for evidence related to each statement. Record your new thinking after the book has been read.

Before		Statement and Evidence	After	
Agree	Disagree		Agree	Disagree
		All of the water on Earth is connected. Evidence:		
		There is water in the atmosphere. Evidence:		
		The water I had this morning may have been used by a dinosaur millions of years ago. Evidence:		
		The amount of water on Earth changes over time. Evidence:		
		Most freshwater is used in homes. Evidence:		
		We use about the same amount of water as people in other parts of the world. Evidence:		
		Our demand for water is growing. Evidence:		

Chapter 14
Let's Dig! Exploring Fossils

OVERVIEW

This inquiry invites students to assume the role of paleontologists in a simulated dig in order to answer the question, *What can we learn from studying fossils?* Correspondence with a mysterious (fictional) paleontologist and the reading of informational text helps students develop a plan for excavating dig sites. As students discover fossils and sediments, they use a variety of texts to answer authentic questions about the fossils and their formation. Finally, they prepare a report about the history of the dig site and the types of living organisms once found there.

The inquiry incorporates a variety of literacy experiences and skills, including shared writing, letter writing, procedural writing, technical writing, and oral presentations. All reading is done on a need-to-know basis and will thus vary from student to student, group to group, and class to class.

OBJECTIVES

- Make inferences about past environments using fossil evidence
- Ask and answer questions about fossils using firsthand evidence and nonfiction text
- Write procedural and informational text
- Write in letter and e-mail formats
- Create drawings and infographics such as maps and diagrams
- Participate productively as a member of a collaborative group
- Collaborate to write a report
- Participate in an oral presentation

STANDARDS ALIGNMENT

National Science Education Standards
SCIENCE AS INQUIRY

- K–4, 5–8 Abilities Necessary to Do Scientific Inquiry

LIFE SCIENCE

- 5–8 Diversity and Adaptations of Organisms

EARTH AND SPACE SCIENCE

- K–4 Properties of Earth Materials
- 5–8 Earth's History

Common Core State Standards for English Language Arts

WRITING

- Grades 3–5 Text Types and Purposes
- Grades 3–5 Production and Distribution of Writing
- Grades 3–5 Research to Build and Present Knowledge

For a detailed standards alignment, see Appendix 3.

TIME FRAME

- Ten 45-minute class periods

SCIENTIFIC BACKGROUND INFORMATION

Fossils are the preserved remains or traces of living organisms. Although many of us associate fossils with dinosaur bones, there are a wide variety of plants, animals, and other microorganisms that have been preserved in the fossil record.

Fossils can be grouped according to the way in which they formed. Four major types are body fossils, impression fossils, mold and cast fossils, and mineral replacement fossils. *Body fossils* include teeth, bones, or entire organisms preserved when trapped in tar, amber, wax, or ice. You may have seen pictures of frozen baby woolly mammoths discovered in Siberia or insects trapped in amber. These are the only type of fossils to include the soft tissues of organisms. Body fossils are quite rare.

Impression fossils show the outline of plants, feathers, fish, or other organisms that are trapped in sediment after they die. Over time, the organism itself decays but leaves a print in the sediment. A dark layer of carbon may remain in the organism's place. Tracks, teeth marks, burrows, and tail marks are also considered impression fossils.

Mold and cast fossils are a type of impression fossils. First, an organism dies and is buried in sediment. It decays, leaving a hole (the mold) in its place. If this mold is later filled with sediment, it produces a three-dimensional model (the cast) that resembles the organism.

Mineral replacement fossils form when an organism is buried in sediment. Water seeping into the bone dissolves the bone, which is replaced by minerals. Many dinosaur bones are examples of mineral replacement fossils. So is petrified wood (Fries-Gaither 2007).

Paleontologists study fossils to learn about the types of organisms that lived long ago and how these organisms are related to those alive today. They also use fossil evidence to reconstruct past climates and conditions. For example, fossilized plants found in South America and Africa support the theory that the two continents were once connected in the supercontinent of Pangaea. Plant fossils found in Antarctica suggest that the continent was once located closer to the equator and had a much warmer climate than today.

When excavating fossils, paleontologists are careful to note the location of each fossil and the characteristics of the sediment surrounding it. By studying the strata, or layers of sediment, scientists can learn more about the environment in which the organisms lived. The law of superposition also helps scientists determine the relative age of the specimens. Fossils found in lower strata are necessarily older than fossils found closer to the surface, unless an earthquake or geologic disturbance has altered the area.

MISCONCEPTIONS

Two general misconceptions about fossils are listed in Table 14.1.

TABLE 14.1. MISCONCEPTIONS ABOUT FOSSILS

Common Misconception	Scientifically Accurate Concept
Fossils are pieces of dead animals and plants.	"Fossils are not actually pieces of dead animals and plants. They are only the impression or cast of the original living thing. The actual living parts decay away but their shape is permanently recorded in the rock as it hardens" (U.S. Geological Survey 2011).
Fossils are rare.	Fossils are plentiful.

Misconceptions about what constitutes a fossil may arise during this unit. During the explore phase of the inquiry, students will be referencing nonfiction texts that include information about fossil formation. As students read about fossil formation, draw their attention to the fact that earth materials replace the original plant or animal parts or that impressions of the organism are left behind. When opportunities arise, ask students what evidence they have to support the texts' explanation that fossils are composed of earth materials rather than actual plant or animal remains.

TEXT SET

Barnum Brown: Dinosaur Hunter by David Sheldon (New York: Walker and Company, 2006); biography, Flesch-Kincaid reading level 6.5.

This book is an illustrated biography of early-twentieth-century paleontologist, Barnum Brown, the man who is credited with first discovering the bones of *Tyrannosaurus Rex*.

Dinosaur Dig! by Susan H. Gray (New York: Scholastic, 2007); reference, Flesch-Kincaid reading level 2.9.

This *Scholastic News* easy reader highlights the process of setting up a dig and recovering dinosaur fossils.

Dinosaur Mountain: Digging Into the Jurassic Age by Deborah Kogan Ray (New York: Farrar, Straus and Giroux, 2010); narrative expository, Flesch-Kincaid reading level 7.0.

The quest to discover dinosaurs in the mid-nineteenth century was one of bitter competition. Learn about the "bone wars" and Earl Douglass's discovery of the first almost complete skeleton of an apatosaurus, one of the largest dinosaurs to ever roam the Earth.

The Dinosaurs of Waterhouse Hawkins by Barbara Kerley (New York: Scholastic Press, 2001); biography, Flesch-Kincaid reading level 5.0.

No one knew what dinosaurs looked like until a Victorian artist named Benjamin Waterhouse Hawkins brought them to life by building life-size models. Learn about Hawkins's life, work, and dedication to his dream of teaching people about these ancient animals.

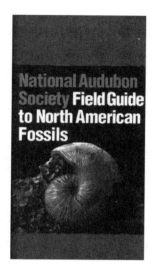

Field Guides. Students will need access to appropriate fossil field guides. For example:

National Audubon Society Field Guide to Fossils: North America by Ida Thompson (New York: Chanticleer Press, 1998); Flesch-Kincaid reading level 7.4.

Fossils by Melissa Stewart (Mankato, MN: Compass Point Books, 2003); reference, Flesch-Kincaid reading level 4.4.

Learn about different types of fossils, where and how they form, and what type of information can be learned from them.

Fossils by Sally M. Walker (Minneapolis, MN: Lerner Publishing Group, 2007); reference, Flesch-Kincaid reading level 4.2.

Fossils are more than just dinosaur bones. Learn about the different types and how they are formed and classified.

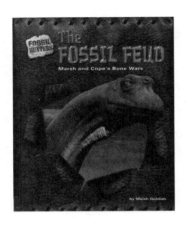

The Fossil Feud: Marsh and Cope's Bone Wars by Meish Goldish (New York: Bearport, 2007); narrative expository, Flesch-Kincaid reading level 4.7.

Meet O. C. Marsh and E. D. Cope, the two main players of the nineteenth century's "Bone Wars." Learn about their competition to find fossils.

Fossils Tell of Long Ago by Aliki (New York: HarperCollins, 1990); reference, Flesch-Kincaid reading level 3.6.

Learn about the different types of fossils, how they were formed, and what we can learn from them. The illustrations have been updated and are now in color, but the text from the original version (published in 1972) remains the same.

Mary Anning and the Sea Dragon by Jeannine Atkins (New York: Farrar, Straus and Giroux, 1999); biography, Flesch-Kincaid reading level 4.6.

A brief biography of Mary Anning, who discovered an ichthyosaur skeleton on the cliffs of Lyme Regis, England, in 1812 when she was just 12 years old. She later became one of the first women paleontologists.

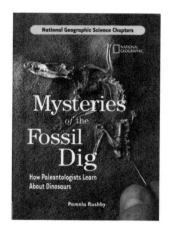

Mysteries of the Fossil Dig: How Paleontologists Learn About Dinosaurs by Pamela Rushby (Washington, DC: National Geographic, 2006); reference, Flesch-Kincaid reading level 5.4.

This book explains how paleontologists dig for and learn about dinosaurs. It also tells the story of how Sue, a *Tyrannosaurus rex* dinosaur, was discovered in South Dakota.

What Do You Know About Fossils? by Suzanne Slade (New York: Rosen, 2008); reference, Flesch-Kincaid reading level 5.1.

This book asks and answers 20 questions about fossils.

MATERIALS

- Aquariums or plastic boxes with lids (underbed storage boxes work well)—one per small group of students
- Sand
- Soil
- Aquarium gravel in a natural color
- A variety of fossils—purchased, from an existing collection, or made according to directions on pages 227–228 (additional materials are listed in those directions)
- String
- Tape
- Small shovels (plastic sandbox shovels are fine)
- Watercolor paintbrushes
- Magnifying glasses
- Trays (Styrofoam trays from the meat department of a grocery store work well)

- Buckets, boxes, or aluminum pans to hold excavated soil
- Colored pencils
- Graph paper
- Document camera (optional)
- Digital camera (optional)
- Copies of supporting documents

SUPPORTING DOCUMENTS

- Letters from Dr. Brown, placed in envelopes and addressed to the teacher
- Site Report Graphic Organizer
- Preliminary Report Outline

SAFETY CONSIDERATIONS

Instruct students in the proper use of digging tools before beginning the excavation. Students should wear protective eyewear during dig sessions and should wash their hands following each dig session.

Let's Dig! Exploring Fossils Inquiry Unit

Advance Preparation: If you do not already have a collection of appropriate fossils, you will first need to create a variety of fossils for use in the dig sites. The instructions given here are from lesson plans available at the Utah Education Network website (2004, 2007). You will need several of each type per small group of students. The following materials are needed to create the fossils: small plastic insects, wax paper, hot glue gun and glue sticks, leaves, plaster of paris, small plastic cups, shells, petroleum jelly, talcum powder, sponges (cut into small pieces), oven-safe containers, sand, salt, water.

- Body fossils: Place small plastic insects a few inches apart on a sheet of wax paper. Use a hot glue gun to completely cover the insects in a mound of hot glue. Let the glue dry, and peel the fossils off of the wax paper. These represent insects that were trapped in amber and preserved.

- Impression fossils: Collect a variety of leaves to be used in creating the fossils. Mix up plaster of paris according to directions on the package. It should have the consistency of a milkshake. Fill plastic cups with about a half inch of plaster of paris. Press leaves, vein sides down, into the plaster. Let these dry overnight, and then remove the leaves and remove the fossils from the plastic cups).

- Mold and cast fossils: Collect a variety of shells to be used in creating the fossils. Mix up plaster of paris according to directions on the package. It should have the consistency of a milkshake. Fill plastic cups with about a half inch of plaster of paris. Cover the shells with a layer of petroleum jelly (so they don't stick to the plaster), then press them about three-quarters of the way into the plaster. When the plaster is almost hard, pull the shells out and leave the plaster to dry completely. Once the plaster is dry, cover the surface with a layer of talcum powder. Mix up more plaster of paris and pour it into the plastic cups, on top of the existing mold. Let this dry completely, then remove the plaster from the cups and carefully separate the two pieces.

- Mineral replacement fossils: Bury at least one small natural sponge per group in an oven-safe container full of sand. There should be a layer of sand below and above the sponges. Next, mix 2 parts salt to 5 parts water in another container, making sure that the salt is dissolved. Carefully pour the salt water on top of the sand until it completely soaks the sand. Place the container in the oven at 250°F for a few hours, until the sand completely dries. (If you do not have a suitable oven-safe container, you can use a regular container and leave it in a warm, dry place until the sand dries. Just know that the process will take longer!) Once the sand is completely dry, remove the sponges. They are ready to be buried in the appropriate layer. The sand can be reused for the dig sites as well.

Next, assemble the dig sites (one per small group of students). In large plastic containers (such as the type used for underbed storage), add layers of materials to represent the various strata. Each layer should be at least an inch thick. Layers will retain their shape best if added in this order:

- Bottom layer: Use sand; bury the mold and cast fossils (keep halves separated) and the sponge mineral replacement fossils in this layer.
- Middle layer: Use aquarium gravel; bury the impression fossils in this layer.
- Top layer: Use soil; bury the body fossils in this layer.

Tamp down each layer firmly before adding the next layer. It is best to place the fossils away from the sides of the box so they are not visible. If using plastic containers, you can place the lid on each to prevent the sites from being disturbed. Label each dig site with a number or letter so that student groups will be able to locate their site easily from day to day.

ENGAGE

In the engage phase, students are invited to participate in a fossil dig. They listen to a read-aloud about Mary Anning and participate in a shared writing experience.

1. Begin by telling students that a mysterious envelope was slid under the classroom door when you came to school this morning. Open the envelope, addressed to your class, and read the letter from Dr. Brown (Letter #1) shown on page 235 out loud. If you have access to a document camera, you might want to display the note and have students follow along as you read.

2. Show students *Mary Anning and the Sea Dragon* and tell them that this book was found with the envelope. Read the book aloud to the class, discussing the story with students and inviting them to share insights and comments as they choose.

3. After reading, ask students to share one thing they liked or learned with a partner. Have a few students share with the entire class. Ask students if they think that they could do what Mary Anning did, and allow them to briefly discuss this in pairs. Again, allow a few students to share their responses.

4. Take a vote (show of hands is fine) to decide whether or not students want to excavate the site. (If students are reluctant to agree, you can give reasons to persuade them or let their enthusiastic peers do the same.)

5. When the class has decided to participate, compose a letter in reply using the shared writing approach. Shared writing is a whole-class activity in which the teacher transcribes the entire

text while engaging students in a thoughtful discussion about the text's content. You might opt to write the letter on chart paper, or on a regular 8½ in. × 11 in. sheet of paper enlarged with a document camera. This shared writing experience is a great time to review letter-writing format. Once the letter is completed, leave it posted in the classroom or place it in an envelope addressed to Dr. Brown. This should conclude the inquiry for the day.

Assess this phase: At this stage of the inquiry, formative assessment should be the focus. Listen to students' discussion about the text and whether or not they think they should take on the challenge of excavating the fossils. This will provide insight into students' prior knowledge of fossils and paleontology, as well as who can become a scientist. Student participation in the shared writing process will help you determine if they are ready to compose letters and messages to Dr. Brown on their own in later phases of the inquiry.

EXPLORE

In the explore phase, students plan and conduct a fossil dig. They analyze the fossils they find and consult informational texts to make inferences about the environment.

1. The next day, tell students that another letter that has arrived. Open the envelope and read the letter from Dr. Brown (Letter #2) shown on page 236.

2. Show students *Mysteries of the Fossil Dig* and tell them that this book was included with the envelope. Read pages 15–22 aloud and discuss with the class.

3. Ask the class to generate a work plan for their dig, rereading parts of the text as needed. Alternatively, you can break students into their small groups and ask each group to generate a plan, using *Mysteries of the Fossil Dig* and the related (but lower-reading-level) *Dinosaur Dig!* for reference.

4. If small groups of students created their own work plans, have each group share their plans with the class and then revise based on feedback. The work plan should be recorded on chart paper and posted in a prominent location. Students should record the work plan in their science journals. It should include:

   - Dividing the dig area into a grid
   - Creating a map of the dig area
   - Digging to locate fossils
   - Mapping the fossils' locations before removing them
   - Carefully removing and cleaning the fossils
   - Drawing and describing the fossils after they are cleaned
   - Taking pictures of the fossils

If students do not include these tasks in their work plan, reread or refer them to the text(s) to find them.

5. Once the work plan is complete, discuss that students will work in small groups and will take turns in the following roles:

    • Excavator: digs carefully until fossils are located

    • Archivist: maps the location and carefully removes the fossil, cleaning it if needed

    • Curator: takes digital pictures, draws, and describes the fossils

6. Students should compose a letter to Dr. Brown outlining their plan. This can be another shared writing experience, or small groups can compose letters. Place these in an envelope, or post the letter(s) prominently. Additionally, post the question, *What can we learn from studying fossils?* in a prominent place in the classroom. This should conclude the day's lesson.

7. The next day, tell students another letter has arrived. Open the envelope and read the letter (Letter #3) shown on page 237.

8. Now it's time to begin the excavation. Give each small group a dig container (aquarium or plastic container) that they will work with for the duration of the explore phase. Introduce students to the sites and review the work plan created by students. At a minimum, the excavation should be conducted over a three-day period, so that one layer is excavated per day. Rotate jobs within each small group so each student has a chance to complete all of the tasks related to the excavation.

9. Before students begin to dig, have them examine the dig site through the side of the box. Ask them to describe what they see, and have them draw a cross section that illustrates this. Written descriptions and cross-sections should be added to their science journals. Remind them that they will be studying and describing the sediment found in each layer as well as the fossils they find. Students should then create a simplified grid and a map of the site. Dividing the container into quadrants with two pieces of string will allow them to map but not impede their ability to dig. Strings can be taped to the outside of the container to hold them in place.

10. As students excavate each layer, they should

    • describe (and name) the sediment(s) found in the layer;

    • plot the location of each fossil on a map of the site;

    • draw and/or photograph, label, measure, and describe each fossil; and

    • identify the fossil, based on field guides.

    • Students can place unwanted sediment into separate containers (buckets, shoe boxes, aluminum foil pans) as they dig. They should take care to not disturb the layers below.

11. At the end of each day's excavation, meet for a brief whole-class discussion. Discuss student

findings and record any questions raised. The following guiding questions can help students focus their attention on important properties of the fossils:

- How are these fossils alike and different?
- How did these fossils form?
- What can we learn from these fossils?
- What does the soil tell us about the fossils in the layer?

Encourage students to turn to texts to find answers to these and other questions they may have raised. Create a basket of reference books and field guides using the text set described earlier in this chapter. Students will add new information to their science journals and write a summary of the findings from that stratum.

If students raise questions that they cannot answer with the texts, remind them that they can share them with Dr. Brown. You can accomplish this in several different ways. You might serve as the go-between, telling students that you will e-mail the questions and then sharing "responses" the next day. Alternatively, you could create a free e-mail account for Dr. Brown, using a service such as Google mail (*http://gmail.com*). You could compose e-mails to the address on your account, either on behalf of the students or as a shared writing experience. If students have access to e-mail, they could compose messages themselves. Later in the day, you can log in to the account and reply to the class or student e-mails. Finally, if you can find a geologist or paleontologist at your local university, you could ask this person to reply to student questions as Dr. Brown. If you do have a real contact, feel free to change the name on the letters!

12. Repeat this process for each layer. After all layers have been excavated, pose guiding questions about the soil:

- Is the soil the same or different throughout?
- Why might there be different layers of soil?

13. Once the excavations are complete, have students compose a message to Dr. Brown explaining that their work is finished and that they'd like to share their findings. In a reply message, Dr. Brown should congratulate students for their hard work and ask them to generate a report (one per group) that explains the types of animals and plants found there, the relative ages of each, and the conclusions they've drawn about the site and its history. Each group will present their findings to the class as well.

Assess this phase: Formative assessment in this phase focuses on process rather than product. If students create work plans in small groups, review these to ensure that they are modeling their plan after the information presented in the texts. If creating the work plan as a whole class, use guiding questions such as "How does your plan compare with what you read in the text?" to keep students focused on applying what they've learned to create the plan. Observe students as they excavate, clean, and document the fossils to ensure that they are following their work plans. Each day's class discussion provides insight into the kinds of observations students are making and questions they are asking about the fossils. Use guiding questions to encourage comparison of the fossils and to help students think about how the fossils formed and the kinds of information they provide. These discussions also may lead you to add books to the basket in order to answer student questions.

EXPLAIN

During the explain phase, students will work in small groups to generate their reports and presentations.

1. Begin by allowing students to finish any classification and research that was not completed during the explore phase.

2. Tell students that they will be creating a report summarizing their findings and that the "Site Report Graphic Organizer" will be a useful tool to organize their thinking. Students should work in their small groups to complete the graphic organizer. During this time, provide support to groups as needed. Students may especially need support as they generate conclusions from the fossil evidence.

3. Gather the whole class together for a brief discussion of the report. Review the required elements (types of fossils, relative ages, and conclusions about the site's history) and ask students to generate a list of components that might be included in a report, based on the graphic organizer. Suggested ideas include diagrams of fossils, a map and cross section of the site showing the locations of fossils and the various strata, and a written description of the group's conclusions. Also discuss the need to use scientific vocabulary, include headings, and write in a formal style. Each group should complete a "Preliminary Report Outline" that details the parts of the report and divides the workload among group members. Review these outlines and meet with groups as needed before allowing students to begin work.

4. Allow each group to select the format of their report and presentation. Possibilities include a traditional paper-and-pencil format, a PowerPoint presentation, a wiki page, a blog post, a video, or a VoiceThread (*http://voicethread.com* [QR Code 1]). Encourage students to think carefully about what format will best fit their purpose.

   QR Code 1

   Students will need several class periods to prepare their reports and presentations. This is a great time to make use of English language arts time, because the report meets many grade-level expectations in terms of writing informational text.

5. Finally, students will present their reports. Invite an adult to view these presentations. The principal could view them and tell students that he or she will share their findings with Dr. Brown, or the community member you identified previously could visit the class on this day. Having an audience besides the class and teacher adds an element of authenticity to the experience.

Assess this phase: The "Site Report Graphic Organizer" and "Preliminary Report Outline" serve as formative assessment. The first provides insight into students' mastery of the concepts and ability to draw conclusions based on the fossil evidence. If students struggle to complete this, guide them to review the information they've recorded in their science journals as a basis for their responses. The "Preliminary Report Outline" ensures that students are organized and that the work is divided among team members. The reports and presentations serve as summative assessment. The "Science and Literacy Rubric" in Appendix 2 can be used to assess student performance, and the "Achievement Grading Standards" (also in Appendix 2) can translate this into a numerical grade. If students fall below 75%, return to the explore phase for additional work before returning to the writing prompt.

EXPAND

In the expand phase, students will learn about the drama and controversy surrounding paleontology in the 19th century.

1. Begin by reading *Dinosaur Mountain* aloud. On a sheet of chart paper, list some of the important figures mentioned in the text: Benjamin Waterhouse Hawkins, Edward Drinker Cope, and Othniel Charles Marsh. Add Barnum Brown to the list. Explain to students that they will read in small groups to learn more about these people.

2. Divide students into small groups—ideally, different groups than those they have worked in previously. Each group will read and discuss a book from the text set: *The Dinosaurs of Waterhouse Hawkins*, *The Fossil Feud*, and *Barnum Brown*. Groups might read and discuss independently, or you might choose to read and discuss each with students during reading centers.

3. After groups have had time to read and discuss, meet as a whole class. Create a chart on the board or on chart paper and ask each group to share what they learned about the person or people discussed in their text. Conclude the inquiry by asking students to share what they've learned about paleontology in the past and present.

Assess this phase: Listen to what students share about the paleontologists they've read about. Student responses, though not written, can serve as a form of summative assessment for the inquiry. The "Science and Literacy Rubric" in Appendix 2 can be used to assess student performance, and the "Achievement Grading Standards" (also in Appendix 2) can translate this into a numerical grade. If students fall below 75%, return to the expand phase for additional work.

REFERENCES

Fries-Gaither, J. 2007. Geologic time, fossils, and archaeology: Content knowledge for teachers. In *Beyond Penguins and Polar Bears*. *http://bit.ly/NFbbLk*

U.S. Geological Survey. 2011. Schoolyard geology: Fossils. *http://education.usgs.gov/lessons/schoolyard/ fossils.html*

Utah Education Network. 2004. TRB 4.4—investigation 2—simulating fossil formations. *www.uen.org/ Lessonplan/preview.cgi?LPid=10288*

Utah Education Network. 2007. Fossil formation fun. *www.uen.org/Lessonplan/preview.cgi?LPid=18978*

LETTER #1

Dear Students:

I am excited to inform you that we have discovered a site near your school that is suspected to hold many different types of fossils. We wish to excavate these fossils and study them. Unfortunately, all of our paleontologists are currently out of the country on their own digs. Would you take on the responsibility of excavating this site and sharing your findings with us? We would greatly appreciate your help.

Sincerely,

Dr. Brown

Director, Institute for Paleontology

P.S. You're probably thinking that you're just kids and can't dig for fossils. But have you heard of Mary Anning? Read this book before you make your decision.

LETTER #2

Dear Students:

Thank you so much for agreeing to help us! I can't begin to tell you how pleased I am. Your findings will help advance our understanding of fossils. The "big" question that you should try to answer in your investigation is, *What can we learn from studying fossils?* I eagerly anticipate your findings and conclusions.

Sincerely,

Dr. Brown

P.S. I almost forgot—you can't just jump in and start digging! You need to have a plan and be very organized as you work. Why don't you read this book and create a work plan to follow?

LETTER #3

Dear Students:

Your plan(s) sounds wonderful! I think you are ready to get to work. Since you can't just leave school and travel to the site, we've arranged for these dig sites to be brought to your classroom. Feel free to send me questions while you work. I'm leaving for a dig myself, but you can reach me by e-mail. Your teacher has my address. Good luck!

Sincerely,

Dr. Brown

P.S. I've sent some other books that I think will help you as you dig. Enjoy!

Name_____ Date_____

SITE REPORT GRAPHIC ORGANIZER

What types of fossils did we find?

List the fossils from youngest to oldest.

What were the sediments like?

LET'S DIG! EXPLORING FOSSILS

Name_____ Date_____

What conclusions can you draw about the site and its history? What evidence do you have for your conclusions?

Evidence	Conclusion

Name_____ Date_____

PRELIMINARY REPORT OUTLINE

Group Members:

_____

_____

_____

Part of Report (Information)	Format	Person Responsible

Chapter 15
Patterns in the Sky

OVERVIEW

The question students will answer through this inquiry is, *What can we discover by observing objects in the sky?* Three separate but related investigations reveal patterns in the sky. The inquiry begins with an investigation into the changing shape of shadows throughout the day and a demonstration that shows the cause of the day/night cycle. In the second and third investigations, students use technology to reveal the phases of the Moon and the motion of constellations.

It's important to keep in mind that students at this age have a very difficult time conceptualizing the Earth-Moon-Sun relationship. Consequently, they shouldn't be expected to explain what causes the phases of the Moon. The goal is for students to recognize that objects in the sky exhibit predictable patterns.

OBJECTIVES

- Observe and explain the changes in shadows throughout the day
- Explain the day/night cycle
- Observe the phases of the Moon
- Observe the motion of constellations
- Generalize from their observations that there is a pattern to the appearance and motion of objects in the sky
- Use technology to collect data
- Collect data over an extended period of time
- Write a dialogue

STANDARDS ALIGNMENT

National Science Education Standards

SCIENCE AS INQUIRY

- K–4 Abilities Necessary to Do Scientific Inquiry
- K–4 Understanding About Scientific Inquiry

EARTH AND SPACE SCIENCE

- K–4 Objects in the Sky

Common Core State Standards for English Language Arts

INFORMATIONAL TEXT

- Grades 3–4 Integration of Knowledge and Ideas

WRITING

- Grades 3–5 Text Types and Purposes

SPEAKING AND LISTENING

- Grades 3–5 Comprehension and Collaboration

For a detailed standards alignment, see Appendix 3 (p. 282).

TIME FRAME

- Nine 45-minute class periods and intermittently as needed for the expand phase.

SCIENTIFIC BACKGROUND INFORMATION

The universe consists of an uncountable number of celestial bodies, all of which are in motion. For much of human history we believed that the Earth was at the center of the motion. Through the careful and sustained observations of Copernicus, Galileo, Tycho Brahe, and many other astronomers, we have come to understand that the Earth, like other celestial bodies, is moving through space in a predictable pattern. With careful and ongoing observations we can gather evidence of the motion of the Earth, Moon, and constellations.

The day/night cycle is the result of the Earth rotating on its axis. Each 24-hour period the Earth makes one complete rotation. As it turns, the portion of the Earth that faces the Sun experiences daylight; the other side of the Earth, the portion turned away from the Sun, experiences the darkness of night. Between daylight and darkness lies twilight, that period of time when dawn is breaking or dusk is falling. The number of daylight hours a particular location experiences depends on its geographic location and the season of the year. Locations along the equator experience nearly equal amounts of daylight and darkness each day and throughout the year. The farther north or south of the equator a location is, the greater the variation in the number of hours of daylight and darkness.

Evidence for the Earth's rotation can be seen in the changes of shadows throughout the day. When the Sun rises in the east, shadows are long and in a westward direction. As the day progresses and the Sun reaches its highest point in the sky, the shadows become shorter until they nearly cease to exist. The Sun then begins to set in the west and the shadows again lengthen, but now they point eastward. We talk about the Sun rising in the east and setting in the west when in fact it is the Earth's rotation from west to east that results in the apparent motion of the Sun.

Another cycle we witness as a result of celestial motion is the phases of the Moon. The phases of the Moon occur due to the Moon's reflection of the Sun's light and its revolution around the

PATTERNS IN THE SKY

Earth. The Moon does not produce its own light. Stars are the only celestial bodies that produce light. All of the other celestial bodies that we see are reflecting starlight. The Moon reflects sunlight.

Figure 15.1 illustrates a view of the Earth-Moon-Sun system from space. In this figure we see that the side of the Moon facing the Sun is always lit, and the side of the Moon turned away from the Sun is always dark. This is true regardless of the relative positions of the Earth and Moon, with the exception of lunar eclipses. The dark side of the Moon is in its self-shadow (Young and Guy 2008). The circle of moons in Figure 15.1 represents the view of the Moon from space throughout the lunar cycle.

FIGURE 15.1. VIEW OF THE EARTH-MOON-SUN SYSTEM FROM SPACE

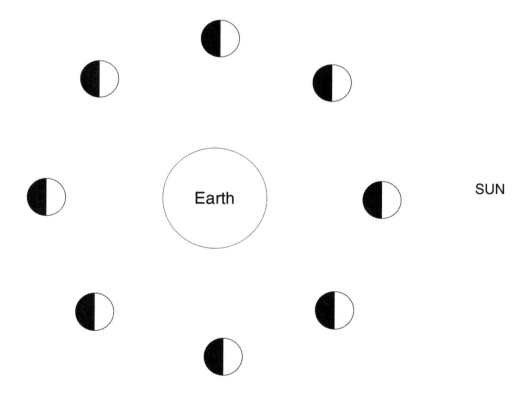

Figure 15.2 (p. 244) illustrates a view of the Moon from Earth. In this figure we see the phases of the Moon as we experience them. When the Moon is on the sunward side of Earth, the portion of the Moon that is reflecting the Sun's light is turned away from us. This is the time of the new Moon, a time when we do not clearly see the Moon in the sky. What we see is the Moon's self-shadow. As the Moon revolves around the Earth, more of the lit side of the Moon becomes visible to us. When the Moon is halfway around the Earth we see the entire lit side of the Moon. This is a full Moon.

FIGURE 15.2. VIEW OF THE MOON FROM EARTH

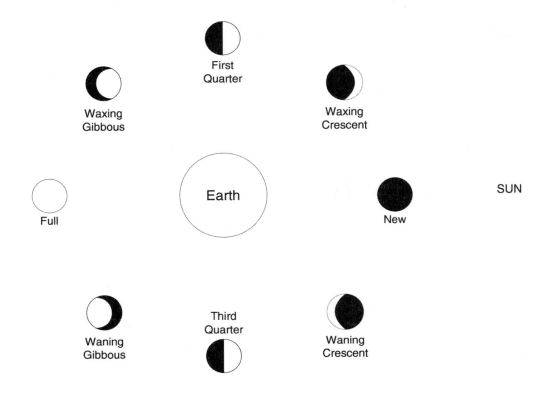

As the Moon travels around the Earth, the amount of the lit side that is visible to us changes. It takes the moon 29.5 days to revolve around the Earth (U.S. Naval Observatory 2011). The circle of moons in Figure 15.2 represents the view of the Moon from Earth throughout the lunar cycle.

Children at this age have a difficult time understanding the Earth-Moon-Sun system. It is sufficient to establish through inquiry that the phases of the Moon occur in a predictable repeated pattern.

The motion of the constellations, as seen from Earth, is due to the Earth's rotation on its axis and revolution around the Sun. As the Earth rotates, the stars seem to rise in the east and set in the west, like the Sun and Moon. In all cases it is the Earth's west-to-east rotation that causes this apparent motion. As the Earth revolves around the Sun, different parts of the sky are visible. This results in seeing different constellations throughout the year. In the Northern Hemisphere we see Orion in the winter but not in the summer, and Ursa Major in the summer but not the winter. Circumpolar stars, such as Polaris, are visible throughout the year (National Research Council Canada 2007).

MISCONCEPTIONS

Throughout human history we have been fascinated by the Moon and stars. Stories about the Moon and constellations are prevalent in folklore and mythology. In spite of our fascination with the Moon and stars, many misconceptions are common about both. The common misconceptions listed in Table 15. 1 come from Operation Physics (1998) and Driver and colleagues (1994).

TABLE 15.1. COMMON MISCONCEPTIONS ABOUT DAY AND NIGHT, THE MOON, AND CONSTELLATIONS

Common Misconception	Scientifically Accurate Concept
It gets dark because: • The Sun goes behind hills • Clouds cover the Sun • The Moon covers the Sun • The Sun goes behind the Earth • The Earth goes around the Sun once a day	The day/night cycle is the result of the Earth turning on its axis once in each 24-hour period.
The shape of the Moon changes because: • Clouds cover part of the Moon • The Moon is in the shadow of a planet • The Moon is in the shadow of the Sun • The Moon is in the shadow of the Earth	Changes in the shape of the Moon are the result of the Moon's reflection of sunlight and its revolution around the Earth.
Stars and constellations appear in the same place in the sky every night.	The locations of stars and constellations seem to change as the Earth rotates on its axis and revolves around the Sun.
The Sun rises in the east and sets in the west every day.	The Earth rotates west to east resulting in the apparent motion of the Sun rising in the east and setting in the west.

Like misconceptions about other science ideas, the misconceptions shown in Table 15.1 may arise from trying to make sense of observations. Asking questions that expose students' thinking is a good way to identify and begin addressing misconceptions. Ongoing questioning and opportunities to respond to prompts in writing allows the teacher to continually monitor student thinking about the concepts. Finally, providing students with multiple and varied ways to interact with the content is helpful in establishing scientifically accurate concepts.

TEXT SET

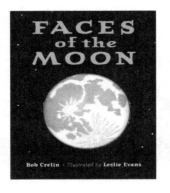

Faces of the Moon by Bob Crelin (Watertown, MA: Charlesbridge, 2009); poetry, Flesch-Kincaid reading level 5.3.

Good information about the Moon's phases is presented in rhyming text, and the illustrations are die-cut to reveal the different phases. Tabs with pictures of the phases make it easy to locate each phase.

Orion by Stephanie True Peters (New York: Rosen, 2003); reference, Flesch-Kincaid reading level 4.7.

This book begins by calling Orion the winter constellation, providing immediate food for thought. The idea that constellations are only visible at certain times of the year is an interesting concept for students to understand. So is the idea that there are supergiant stars, hundreds of times larger than our Sun, and Orion has two of them.

Starry Messenger: Galileo Galilei by Peter Sís (New York: Farrar, Straus and Giroux, 2000); biography, Flesch-Kincaid reading level 4.8.

Most obviously, this book chronicles the the life of Galileo and his contributions to science. But it also tells what can happen when scientific and traditional theories conflict. The artwork is intriguing, and most of it is in the style of Galileo's day, but have a look at the endpapers—they are fun and interesting to compare.

The Big Dipper by Stephanie True Peters (New York: Rosen, 2003); reference, Flesch-Kincaid reading level 4.8.

Two "Dippers" for the price of one! This book describes the Big Dipper, recounts ancient legends about it, and identifies the individual stars that compose it. It also discusses the Little Dipper and the North Star.

MATERIALS

For the Class
- Chart paper
- Tape
- Sunrise and sunset images
- Globe
- Container for drawing 3" × 5" date cards from (e.g., shoebox or plastic container)

For Each Group
- Sidewalk chalk
- Clipboard
- Flashlights
- Balls (basketball size)
- Clay
- Golf tees
- Drawing supplies (paper, markers, crayons)
- 3" × 5" cards
- Computer with internet connection
- Copies of supporting documents

SUPPORTING DOCUMENTS

- "Shadow Tracker" data collection sheet
- "Evidence Tracker" data collection sheet
- "Me and My Shadow" (Keeley, Eberle, and Dorsey 2008)
- "Darkness at Night" (Keeley, Eberle, and Tugel 2007)
- "Constellation Tracker" data collection sheet

SAFETY CONSIDERATIONS

Make sure that students know they should never look directly at the Sun.

Patterns in the Sky Inquiry Unit

ENGAGE

In the engage phase students learn about Galileo and share their thinking about day and night, the Moon, and constellations.

1. Ask students how they are today. Talk about what you did last night or ask what they did last night. Lead a discussion about how often we use the terms *day* and *night* in conversation and media. Give examples like "in the news today," "today's weather," and "at dinner last night." Then ask students what we mean by *day* and *night*. Distinguish between *day* as the time between dawn and dusk and *day* as a 24-hour period.

2. Read aloud *Starry Messenger* to the class. Discuss the methods Galileo used to learn more about the solar system and our Moon. Share that Galileo made observations over long periods of time using only a telescope.

3. Ask students to write and illustrate a paragraph in which they share their current thinking about shadows, day and night, Moon phases, and constellations. Introduce the writing by sharing some of your observations and what you have wondered about the same topics. Something similar to "I've never used a telescope, but like Galileo I've noticed some things about objects in the sky. One day I noticed that my shadow was different at lunchtime than in the morning. Another time I noticed that the Moon seems to change its shape and that the constellations, those groups of stars, seem to move around in the sky. Noticing these things made me wonder what causes the differences. I'm also curious about why we have day and night. Maybe you can help me figure all of this out. What do you know about shadows, the shape of the Moon, constellations, and why we have day and night?" Ask students to help you by sharing what they know about shadows, day and night, the shape of the Moon, and constellations. Collect the illustrated paragraphs and set them aside. Students will use them again later in the inquiry.

4. Share with students that now we have advanced technological tools that allow us to examine data that has been collected over long periods of time, and we can use that data to predict many things about objects in the sky. Introduce the question, *What can we discover by observing objects in the sky?*

Assess this phase: The illustrated paragraphs written by students serve as formative assessment for this phase of the inquiry. They provide a look at students' current understanding of shadows, day and night, Moon phases, and constellations. Consider student responses in light of the misconceptions described in Table 15.1 and how you might modify instruction in response to them. Do not correct misconceptions or incorrect ideas at this point. Students will return to these paragraphs later in the unit as a form of self-assessment and reflection.

EXPLORE: DAY AND NIGHT

In the first explore phase students conduct a firsthand investigation to determine the cause of day and night. At this age students can grasp the concept of the Earth spinning on its axis as the cause of day and night.

Advance preparation: Draw a simple map of the school grounds on the "Shadow Tracker" data collection sheet. Include major features such as streets and the school. Indicate "north" on the map. Mapping programs like Google Maps and MapQuest may be helpful.

Shadow Tracker

1. Begin this activity as early as possible in the day. Organize the class into groups of three or four students. Students are likely to have had experiences with shadows and will know that if they change their position, the appearance of their shadow will also change. It is still worthwhile to reinforce this with a quick game of shadow tag. Shadow tag is an adaptation of tag in which students tag one another by stepping on their shadows.

2. After students have played shadow tag for a few minutes, review the nature of shadows with them. The goals of reviewing the nature of shadows are to have students articulate that a shadow occurs when an object blocks light and to introduce the concept of a self-shadow. The following questions may be helpful (answers are in parentheses):

   • How do shadows form? (An object blocks the path of the light.)

   • What can you do to change the shape of a shadow? (Move or reposition the object that is blocking the light. Demonstrate that the shape of your shadow changes when you move.)

   • What do you notice about the side of the object that is turned away from the light source, in this case the Sun? (The side of the object that is turned away from the light source is dark. Confirm by examining the self-shadow of several objects. This is important later in the inquiry.)

3. At this point tell students that an object that is blocking light forms two kinds of shadows: the cast shadow, the one they used when playing shadow tag; and the self-shadow, the one that makes the object blocking the light appear dark on the side turned away from the light. Figure 15.3 illustrates cast shadows and self-shadows. To illustrate the self-shadow, have the students turn so they are facing the Sun (they should not look into the Sun) and ask them to notice how bright their clothes look. Now have the students turn their back to the Sun, and ask them to notice how their clothes are now darker. The darkness of their clothes is due to the self-shadow.

FIGURE 15.3. CAST AND SELF-SHADOW

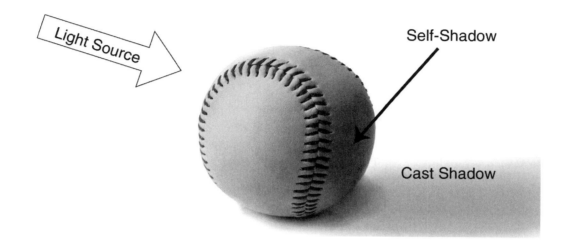

4. Next, give each group sidewalk chalk, a clipboard, and a simple map of the school grounds (use the "Shadow Tracker" data collection sheet). Review the map to help orient students to the cardinal directions (N, S, E, W). Model for them how they will:

    a. Select a stationary object (e.g., a lamppost or a sign) on the school grounds that casts a shadow over pavement, sidewalk, or another surface that can be drawn on with the sidewalk chalk.

    b. Outline the shadow with sidewalk chalk, record the time of the observation, and sketch the outline and object on their "Shadow Tracker" sheets.

    c. Indicate the position of the Sun on the map.

    Alternatively each group could select one student to cast the shadow. In this case it is important that the student stand in the same location each time. To accomplish this, ask the student to stand still while another student from the group first outlines his or her feet and then outlines the standing student's shadow. For the next two observations, the standing student should position his or her feet inside the chalk outline of the feet from the first observation. Have the group indicate on the map where the student is standing and the position of the Sun.

5. Return to the classroom and give each student a copy of the "Me and My Shadow" assessment probe. Read the probe aloud to the class, making sure the students understand that the probe is asking what will happen to the length of the shadow from the time the Sun rises until it sets. Ask each student to respond to the probe independently. Collect the probes and review them before the end of the day to reveal students' early thinking about shadows.

6. Repeat the shadow pattern observation two more times, once near noon and again just before the end of the school day. Each time, students should document their observations on the "Shadow Tracker." After the final observation, return the "Me and My Shadow" assessment probe. Ask students if, based on the evidence they now have, they agree with their earlier thinking. Organize students into pairs or small groups to discuss their early thinking, the evidence they gathered, and their current thinking. If students now have different ideas, they should record their new ideas beneath their initial thinking. Collect the responses and review them before the next day.

7. The next day project the "Shadow Tracker" map or draw it on the board or chart paper. Ask each group to indicate on the map where they made their observations. Students will share their sketches by taping their data collection sheets side by side on the board. Students will compare the sketches, looking for similarities in the way the shadows changed throughout the day.

8. As the students are comparing the sketches, ask guiding questions to focus their thinking,
   • What do you notice about the shadows throughout the day? (They change size, shape, and direction.)
   • How do these changes compare for different items at the same time of day? (The changes in the shadows are all similar.)

9. Give each student an "Evidence Tracker" data collection sheet. As a class, summarize the observations made about the changes of the shadows in one sentence. Example: "The shadows changed throughout the day." Write the sentence on chart paper. Each student should also record the sentence on the "Evidence Tracker" on the first line of the "Shadow Tracker" row. Post the chart paper in an easily accessible place in the classroom.

10. Draw students' attention to their record of the position of the Sun throughout the day. Ask how the position of the Sun in the morning compares with the position of the Sun later in the day. Students are likely to say that the position of the Sun changed. Wonder aloud about what changed the position of the Sun or the motion of the Earth.
    • How can you explain these changes?
      » You moved to change the shape of your shadow in shadow tag. Do the objects move throughout the day?
      » If the objects didn't move, how did the shape of the shadows change?
    • In what direction does the Sun seem to move? (east to west)

11. As a class, summarize the observations made about the position of the Sun throughout the day. Example: "The sun seemed to move from east to west." Write the sentence on chart paper. Each student should also record the sentence on the "Evidence Tracker" on the second line of the "Shadow Tracker" row.

Sunrise, Sunset

12. Have students return to their small groups. Once students are settled in their groups pose the following questions.

    • Where was the Sun early in the morning, long before we started outlining the shadows?

    • Where did the Sun go in the evening, long after we were finished tracking the shadows?

    You may want to write the questions on the board for students to reference as they work.

13. Students should discuss the questions, using their "Shadow Tracker" data collection sheet for reference. Each group should come up with a statement that explains their thinking. As the students discuss the questions, circulate through the room listening to their conversation and providing assistance as needed. Remind students that they should use the evidence about the Sun's position throughout the day to help them come up with a statement.

14. When all of the groups have a statement ready, they will share them with the rest of the class. One member from each group should read their statement. Additional group members can help explain the group's reasoning, linking their thinking to the evidence they collected the day before. Help students recognize that the Sun followed a particular path as the day progressed and was most likely farther to the east in the early morning and farther to the west in the evening.

15. Inform the class that since the Sun comes up before they are in class and sets after they leave they are going to gather evidence about where the Sun is early in the morning and late in the evening by looking at pictures. Show the class some images of sunrise and sunset. These images could come from a variety of sources such as magazines, internet, calendars, photographs, or art. Discuss sunrise and sunset. Ask questions that help students focus on sunrise and sunset, what the sky looks like, where the Sun is, whether the Sun comes up slowly or just appears in the sky all at once, and whether it gets dark all at once in the evening or gradually. Help students see that the Sun seems to come up over the horizon in the morning and sink below the horizon at night.

16. As a class, summarize the observations made about where we see the Sun in the early morning and later in the evening. Example: "The Sun comes up over the horizon in the morning. The Sun sinks below the horizon in the evening." Write the sentence(s) on chart paper. Each student should also record the sentence(s) on his or her "Evidence Tracker" in the "Sunrise, Sunset" row.

17. Give each student the "Darkness at Night" assessment probe, making sure students understand that by night you mean the period between sunrise and sunset. Ask students to respond to the probe. Collect the probes and review them before the next activity. This will provide some insight into what the students already know about the day/night cycle.

Day and Night

18. Begin the following day by summarizing the observations students have made so far:
    - Shadows change shape throughout the day.
    - The Sun seems to move across the sky from east to west throughout the day.
    - The Sun comes up over the horizon in the east in the morning.
    - The Sun sinks below the horizon in the west at night.

19. Set a globe and bright flashlight where all of the students can see them. Share with students that you were thinking that you could use the globe and flashlight to explain the observations they have made about the Sun and shadows.

20. Find your location on the globe. Use this as the starting point for the demonstration.

21. Stand far enough away from the globe so that the flashlight illuminates as much of the side of the globe facing you as possible. Point the flashlight toward the equator (see Figure 15.4). Make sure students understand that the flashlight represents the Sun.

FIGURE 15.4. RELATIVE POSITIONS OF FLASHLIGHT AND GLOBE

22. Ask a student to slowly turn the globe counterclockwise while you hold the flashlight. Turn the lights off and repeat the process. You may have to repeat the demonstration so that all of the students can see both sides of the globe equally well. Consider asking half of the students to stand with you and the other half to stand on the opposite side of the globe. Then have the students switch sides and repeat the demonstration. After the demonstration ask students what they noticed about the portion of the globe facing the flashlight and what they noticed about the portion of the globe that was on the opposite side of the flashlight.

23. Ask the class what they noticed about the amount of light their location on the globe received as the globe turned on its axis. After students have shared their observations ask them to connect what they observed in the demonstration to the observations they reviewed earlier. Follow up with an explanation of the Earth spinning on its axis as the cause of day and night. Include in the explanation that it takes 24 hours, or one day, for the Earth to make one complete rotation.

24. As a class, summarize the observations made about the Earth spinning on its axis. Example: "Day and night are the results of Earth spinning on its axis." Write the sentence on chart paper. Each student should also record the sentence on the "Evidence Tracker" on the first line of the "Day and Night" row.

25. Have students work in their groups to link their new understanding of the Earth spinning on its axis to explain the changing shape of the shadows. Give each group a ball (basketball size), a flashlight, a small piece of clay, and a golf tee. Show students how to use the clay to stick the golf tee to the ball. Challenge students to use the flashlight and ball to demonstrate how the shadows change as the Earth spins. As students experiment, circulate through the groups asking questions and offering assistance when needed. After the students have finished experimenting, ask them to share their results.

26. As a class, summarize the observations made about how the Earth spinning on its axis influences the shape of shadows. Example: "Shadows change shape as the Earth spins on its axis." Write the sentence on chart paper. Each student should also record the sentence on the "Evidence Tracker" on the second line of the "Day and Night" row.

Assess this phase: Formative assessment is used throughout this phase. Student discussion of the nature of shadows following the shadow tag game can be used to informally assess their level of understanding. If needed, spend time developing the concepts of how shadows are formed and of a self-shadow before tracking shadows with sidewalk chalk. The "Me and My Shadow" and "Darkness and Night" assessment probes, the "Evidence Trackers" data collection sheets, and the discussions all provide evidence of students' developing understanding. If students struggle, pose additional guiding questions to help them interpret the evidence that they have collected.

Explain: Day and Night

1. Return the "Me and My Shadow" and "Darkness at Night" assessment probes to the students. Students will review and, if necessary, revise their previous responses.

2. After they have finished, students will write a dialogue modeled after the probe in which a group of friends are discussing why shadows change throughout the day or why it gets dark at night.

Assess this phase: The dialogues serve as summative assessment for the "Day and Night" part of this inquiry. The "Science and Literacy Rubric" in Appendix 2 (p. 276) can be used to assess student performance, and the "Achievement Grading Standards" (also in Appendix 2, p. 277) can translate this into a numerical grade. If students fall below 75%, return to the explore phase for additional work before returning to the writing prompt.

EXPLORE: MOON PHASES

In this second explore phase, students use online data and images to learn about the phases of the Moon. Explaining the cause for the changing phases of the Moon is conceptually too difficult for most students at this age. The emphasis should be on the predictable pattern of the Moon phases.

Advance preparation: Write the dates for the next 30 days on separate 3 in. × 5 in. cards. Put them in a container (e.g., shoe box or plastic bowl) with the dates down so they are not visible. On a second set of 3" × 5" cards, write the text from each of the tabbed pages, beginning with the new Moon, in *Faces of the Moon.* See the example below. Put this set of cards aside; do not put them in the container with the date cards. This second set of cards will be used for the Reader's Theater. Here is an example of Reader's Theater card text:

> The Moon's first phase, we call it NEW,
> when Moon's between the Sun and you.
> Her sunlit side is turned away,
> and we can't see her, night or day.

1. Create a transition by stating, "We've uncovered the pattern of motion that results in day and night. Let's expand our thinking to include the Moon" or something similar.

2. Ask students to draw a picture of the Moon. Hang the drawings in an easily visible location in the room. Compare drawings. Point out that they do not all look the same. (If they do all look the same, see step 3.) Wonder aloud if we can predict what the Moon will look like in a week or two.

3. If all of the drawings look the same, ask students if they have ever noticed that the Moon sometimes looks different. Ask students to share what they have noticed. Wonder aloud if we can predict what the Moon will look like in a week or two.

4. Go to *http://aa.usno.navy.mil/imagery/moon* (QR Code 1) to find out what the Moon looks like today. Ask students to describe the Moon, and ask if it reminds them of anything. QR codes for the U.S. Naval Observatory and AstroViewer links in this unit have been provided for your convenience; use a scanning app on your smartphone, on your tablet, or with the webcam on your computer to scan and quickly access the videos.

5. Ask students to check the Moon tonight to see if it looks like the program predicted it would. It may be necessary to skip this step if the Moon rises after students are likely to be asleep. Go to *http://aa.usno.navy.mil/data/docs/RS_OneDay.php* (QR Code 2) to check the time the Moon will rise at your location.

6. The next day confirm that the Moon looked as predicted.

7. Return to *http://tycho.usno.navy.mil/vphase.html* (QR Code 3) to find out what the Moon will look like tonight (the second night in a row). Then ask the students if they would like to find out what the Moon will look like every night for the next month. Show the class how you can select the dates to see what the Moon will look like at any time in the future or the past. Have each student draw a date card from the container, enter the date, find out what the Moon will look like on that day, and draw what he or she sees on the side of the card without the date. When all of the students are finished, ask them to work together to arrange the cards with shapes of the Moon in a logical order.

8. After they are finished arranging the cards, discuss the arrangement. Ask students to explain their reasoning. To confirm if their arrangement is correct, turn the cards over to reveal the dates. If necessary rearrange the cards in the correct order based on date. Turn the cards over again and compare the correct order with their original sequence. Hang the cards in the correct order on a wall in a prominent location.

9. Based on the correct sequence, ask students to predict what the Moon will look like for the next week in the sequence. Go back to *http://tycho.usno.navy.mil/vphase.html* to find out what the Moon will actually look like on those dates. Compare their predictions with the images on the website. Ask students to volunteer to check the Moon on each of the dates to confirm that it actually looks like they expect it to.

10. To reinforce the predictable pattern of the Moon phases, read *Faces of the Moon*. There are several places in the book where the author describes the Moon as shadowed. This first appears on page 3:

> Our mighty Sun, with sunbeams bright, shines upon the moon in flight and pains her shadowed face with light to make the phase we see.

11. Stop at this point and remind the students about the self-shadows they noticed while tracking shadows. Explain that the shadow mentioned through the book is the Moon's self-shadow.

12. As a class, summarize the observations made about how the shape of the Moon seems to change. Example: "The shape of the moon changes in a predictable way." Write the sentence on chart paper. Each student should also record the sentence on the "Evidence Tracker" in the row labeled "Moon Phases."

Assess this phase: Formative assessment in this phase includes students' drawings of the Moon, discussion around the Moon's various shapes, students' ability to correctly predict the shape of the Moon, and students' ability to arrange the images of the Moon in the correct order. If students struggle with these activities, examine the Moon's shape on additional dates and use guiding questions to help students interpret this evidence.

EXPLAIN: MOON PHASES

In the explain phase students will take turns reading a page of *Faces of the Moon.* Using a Reader's Theater approach, the reader will use voice, facial expression, and gestures to help the rest of the class imagine the phase of the Moon he or she is describing.

1. Divide students into teams of two. Assign each team a phase of the Moon described on one of the tabbed pages in the book. Give the team the corresponding 3" × 5" card with the text from the book. The team will practice reading the text until they are fluent. As they practice they will decide the best way to use their voice, facial expressions, and gestures to bring the text to life. After students are finished rehearsing, call on them in random order to read their cards. When all of the cards have been read, the class will work together to arrange the teams in the correct order based on the Moon phase they read about.

2. Next, have each student pick a future date that is important to him or her, such as the student's birthday, a best friend or sibling's birthday, or a favorite holiday. After choosing a date return to *http://tycho.usno.navy.mil/vphase.html* and find out what the Moon will look like on that day. Then, using the Moon phases sequence cards for reference, draw what the Moon will look like for a week before and a week after the date.

Assess this phase: Students' descriptions of the Moon before and after the selected date serve as summative assessment for the "Moon Phases" part of the inquiry. The "Science and Literacy Rubric" in Appendix 2 (p. 276) can be used to assess student performance, and the "Achievement Grading Standards" (also in Appendix 2, p. 277) can translate this into a numerical grade. If students fall below 75%, return to the explore phase for additional work before returning to the writing prompt.

EXPLORE: CONSTELLATIONS

In the third explore phase, students conduct an online investigation into constellations.

QR Code 4

Advance preparation: Go to *www.astroviewer.com/interactive-night-sky-map.php* (QR Code 4) and click the "Start AstroViewer" button on the left side of the screen. The interactive AstroViewer map will open in a new window. Become familiar with the program before introducing it to students. On the day you are going to use AstroViewer, set it up on each computer the students will be using. Select the location and set the time by clicking on the "Now" button.

1. Begin by asking, "Besides the Moon, what else do we see in the nighttime sky?" (stars)

2. Show the class the cover of *The Big Dipper.* Discuss the cover, then read pages 5, 13, 14, and 18, pausing to ask questions and point out important details in the text and the images. Share with students that they are going to use AstroViewer to learn more about the Big Dipper and constellations.

3. Demonstrate how to use AstroViewer by selecting the location and clicking the "Now" button. Point out some of the constellations. Click the hours, days, months, and years arrows (>) to demonstrate how the view of the sky changes.

    a. Find Polaris. Confirm that it is stationary and located in the north.

    b. Locate the Big Dipper. Click through month by month to see how the asterism moves throughout the year.

    c. Click the "Now" button and point out some of the constellations that can be observed in the sky at the time of the lesson.

    d. Pick a constellation and demonstrate how you can follow it as it moves across the sky throughout the year by clicking the hours, days, and months arrows.

4. Read pages 5 and 22 of *Orion.* Use AstroViewer to illustrate how Orion's position changes throughout the year.

5. Pair students up in teams of two, and ask them to select a constellation to follow. Each team should then go to AstroViewer, locate their constellation, and click through month by month to see when and where the constellation is visible. Students will record the position of the constellation each month on the "Constellation Tracker" data collection sheet. When the constellation is below the horizon, the students should leave the diagram for that month blank.

6. After teams have recorded the position of their constellation, they will share their data collection sheets with the class. Tape the data collection sheets to the wall and compare them. Ask students what they notice about the position of the constellations throughout the year. They should notice that the constellations seem to move in the nighttime sky throughout the year.

7. As a class, summarize the observations made about how the position of the constellations seems to change. Write the sentence on chart paper. Each student should also record the sentence on the "Evidence Tracker" in the row labeled "Constellation Tracker."

8. Have students try to find the Big Dipper and their constellation (if it is visible at the time) in the nighttime sky.

Assess this phase: Students' observations about the constellation data serve as formative assessment for this phase of the inquiry. If necessary, use guiding questions like "How does your constellation's position change over the course of a year?" and "Do all the constellations' positions change in the same way?" to help students interpret the data to identify the pattern.

EXPLAIN

In this final explain phase students review their evidence and formulate an answer to the question, *What can we discover by observing objects in the sky?*

1. Review the set of summary statements listed on the chart paper that correspond to the student "Evidence Tracker" sheet. At this point there should be eight statements. Ask students what they notice about the statements. They should notice that all of the statements involve some kind of pattern: a pattern of shadows changing shapes, a pattern of the Sun's position in the sky, a pattern of day and night, a pattern to the phases of the Moon, and a pattern to the positions of the constellations in the nighttime sky.

2. Have students write and illustrate a paragraph in which they share their current thinking about shadows, day and night, Moon phases, and constellations. After they have finished writing their paragraphs, hand back the paragraphs they wrote at the beginning of the inquiry. In a final paragraph students will reflect on how their thinking has changed since they began observing objects in the sky.

Assess this phase: Paragraphs serve as the summative evaluation for the inquiry as a whole. The "Science and Literacy Rubric" in Appendix 2 (p. 276) can be used to assess student performance, and the "Achievement Grading Standards" (also in Appendix 2, p. 277) can translate this into a numerical grade. If students fall below 75%, return to the explore phase for additional work before returning to the writing prompt.

EXPAND

In the expand phase, students make firsthand observations of celestial motion over time.

Refer back to Galileo and the methods he used to learn about the motion of objects in the sky. Compare the tools Galileo used with the tools the students used. Have the students devise a plan to confirm what they have learned through simulations by making observations of the moon, constellations, and day length. Here are some suggestions:

- Record the time of the sunrise and sunset for two weeks. Calculate the day length for each day. Prepare a graph that illustrates the day length trend for a particular location. Predict the time the Sun will rise and set for the next three days.

- Observe and draw the phases of the Moon through one complete moon cycle. Also record the time of day and the location of the Moon in the sky. At the beginning of the next Moon cycle, predict what the shape of the Moon will be after 7, 14, and 21 days. Observe the Moon on these days to confirm the accuracy of the predictions.

- Select a constellation to follow throughout the school year. Look for the constellation once a week and record the position in the sky where the constellation was observed. Compare observations with expected locations.

It is not necessary that every student make every observation. Consider allowing students to follow their interests and share from time to time what they have learned.

Assess this phase: Assessment focuses on student observations of the Sun, Moon, and/or constellations. This is a time to assess science process skills: prediction, observation, and data collection. Students are not only expanding their understanding of patterns but also developing their expertise with these process skills. Assessment should be focused not on a correct answer but on the process skills themselves. We recommend using the "Science Process Skills Rubric" in Appendix 2 (p. 277).

REFERENCES

Driver, R., A. Squires, P. Rushworth, and V. Wood-Robinson. 1994. *Making sense of secondary science: Research into children's ideas*. New York: Routledge.

Keeley, P., F. Eberle, and C. Dorsey. 2008. *Uncovering student ideas in science, vol. 3: Another 25 formative assessment probes.* Arlington, VA: NSTA Press.

Keeley, P., F. Eberle, and J. Tugel. 2007. *Uncovering student ideas in science, vol. 2: 25 more formative assessment probes.* Arlington, VA: NSTA Press.

Operation Physics. American Institute of Physics. 1998. Children's misconceptions about science. American Institute of Physics. *www.eskimo.com/~billb/miscon/opphys.html*

National Research Council Canada. 2007. Motions in the sky. *www.nrc-cnrc.gc.ca/eng/education/astronomy/topics/motions.html*

U.S. Naval Observatory. 2011. Phases of the Moon and percent of the Moon illuminated. *http://aa.usno.navy.mil/faq/docs/moon_phases.php*

Young, T., and M. Guy. 2008. The Moon's phases and the self shadow. *Science and Children* 46 (1): 30–35.

Name_____ Date_____

SHADOW TRACKER

Place an X on the map at the location of the object you are observing. Mark the position of the Sun by writing "Sun" on the map.

Time:

Time:

Time:

Map

Name_____ Date_____

EVIDENCE TRACKER

Activity	Important Observations
Shadow Tracker	
Sunrise, Sunset	
Day and Night	
Moon Phases	
Constellation Tracker	

Look over the observations. What did you discover by observing objects in the sky?

_____

_____

_____

_____

_____

Name_____ Date_____

Me and My Shadow

Five friends were looking at their shadows early one morning. They wondered what their shadows would look like by the end of the day. This is what they said:

Jamal: "My shadow will keep getting longer throughout the day."

Morrie: "My shadow will keep getting shorter throughout the day."

Amy: "My shadow will keep getting longer until it reaches its longest point and then it will start getting shorter."

Fabian: "My shadow will keep getting shorter until noon and then it will start getting longer."

Penelope: "My shadow will stay about the same from morning to day's end."

Which friend do you most agree with? _____

Describe your thinking. Explain the reason for your answer.

_____

_____

_____

_____

_____

Source: Keeley, P., F. Eberle, and C. Dorsey. 2008. *Uncovering student ideas in science, vol. 3: Another 25 formative assessment probes.* Arlington, VA: NSTA Press.

Name_____ Date_____

Darkness at Night

Six friends were wondering why the sky is dark at night. This is what they said:

Jeb: "The clouds come in at night and cover the Sun."

Talia: "The Earth spins completely around once a day."

Nick: "The Sun moves around the Earth once a day."

Becca: "The Earth moves around the Sun once a day."

Latisha: "The Sun moves underneath the Earth at night."

Yolanda: "The Sun stops shining."

Which friend do you think has the best reason for why the sky is dark at night? Describe your ideas about why the Earth is dark at night and light during the day.

_____

_____

_____

_____

_____

Source: Keeley, P., F. Eberle, and J. Tugel. 2007. *Uncovering student ideas in science, vol. 2: 25 more formative assessment probes.* Arlington, VA: NSTA Press.

Name_____ Date_____

CONSTELLATION TRACKER
Constellation: _____

January	February	March
April	May	June
July	August	September
October	November	December

Chapter 16
Final Thoughts

We have thoroughly enjoyed sharing with you what we believe is an important instructional practice—uniting science and literacy in an authentic way that nurtures student development in both disciplines. We hope that this approach will ignite in you a desire to adopt the practice of including multigenre nonfiction text sets into science instruction as the norm in your classroom. Additionally, we hope our enthusiasm is evident from these chapters and contagious to the point that you will embrace the challenge of developing your own units. When you do, you'll find that the way you think about science and literacy instruction will change. You will begin to see that the goals of each are not only complementary, but often the same. Here are some things to think about as you begin to develop units.

WHERE TO BEGIN

Do you begin with the topic or a really great book? When we started, we wanted to begin with really great books and write units around them. This occasionally worked, but we found that if we were going to develop standards-aligned units more often than not we had to start with the topic. We also found that starting with the topic led us to some really great books.

SELECTING TEXTS

How do you find the right books? You would think that finding the right books requires extensive knowledge of what is available in children's literature. In reality it requires spending time in the children's section of your local library. After all, your goal is to find the right books for your classroom, and these need to be easily available! We found ourselves pulling dozens of books off the shelves for each topic. We scanned and rejected most, checked out the remainder for further consideration, and narrowed these down to a manageable number. Finding the right books is critical. Here is a list of questions we often asked ourselves as we reviewed books:

- Is the content scientifically accurate?
- Does the book avoid anthropomorphism?
- Is the book free of gender, ethnic, and socioeconomic bias?
- Would the book be interesting to most students and require the active participation of students in their own learning?
- Does the book frame the content in a way that is meaningful to students?
- Would reading the book lead to significant conversation and reflection about scientific concepts and processes by the students and teacher?

- Are the illustrations and models used in a highly effective way that supports the content?
- Is the book clearly written and carefully edited?
- Can the book be used in a variety of ways?

The questions that most often resulted in a book being rejected were scientific accuracy and anthropomorphism. It can be challenging to evaluate the scientific accuracy for a topic with which you are not that familiar. We researched topics to make sure we had the most accurate scientific information and often discussed books with each other to determine if language or illustrations were inaccurate or misleading. Scientific inaccuracies are sometimes subtle. When this is the case it is very tempting to use the book anyway. Our worry in using a book that contains scientific inaccuracies is that the inaccuracy will support a misconception that will be difficult or impossible to remedy later.

Anthropomorphism is very common in children's literature. We are not particularly concerned about anthropomorphism in children's fiction. But it can be very problematic when the texts are used in the teaching of science. Children in grades 3–5 most likely know that animals do not talk, dress up in clothes, and so on. But more subtle anthropomorphism can lead to misconceptions. For example, if the text implies in some way that an organism chooses to have a particular trait or exhibit an instinctive behavior, students may then have a difficult time understanding the nature of adaptations. Also watch for anthropomorphism as it relates to plants. We have seen children's books that refer to seeds sleeping or plants eating and drinking. Again, this use of language can lead to the formation of misconceptions.

PUTTING IT ALL TOGETHER

What's next? Once you have chosen a topic and selected the books, you'll need to decide on a question students can investigate and outline how they will conduct the investigation (unless you are planning an open inquiry unit). The question must be one that requires an evidence-based answer. The evidence can come from first- and secondhand investigations. It is important to remember that if the unit is to be an inquiry unit, the students must collect evidence and make evidence-based claims. Otherwise the students are doing hands-on rather than minds-on science. From our perspective, the learning cycle is the absolute best way to structure an inquiry unit. The learning cycle framework we've used includes the exploration phase during which students collect data and the explanation phase in which they use the data to make evidence-based claims and explain their findings. The version we've used also includes the equally important engage and expand phases, and supports assessment throughout.

We hope you have enjoyed this book, had success implementing the units with your students, and are ready to develop your own. Good luck and happy inquiring!

Appendix 1

Background Information: Science Process Skills and Literacy Strategies and Techniques

Throughout our inquiry units, we have referenced and included science process skills and a variety of strategies and techniques intended to support students' literacy development. Here, we provide background information on these skills, strategies, and techniques.

SCIENCE PROCESS SKILLS

These process skills have been described many places. We referenced Padilla (1990) and Rezba, Sprague, and Fiel (2003) for our descriptions.

BASIC SKILLS

- *Observing* is gathering information about an object or phenomena. Observations may be qualitative or quantitative. We make qualitative observations with our senses. We might notice the color and shape of an object, its texture, smell, or the sound it makes when shaken.

- *Inferring* is a thought process that results in an explanation based on previous knowledge and new observations. If I see my friend come inside with an umbrella, I might infer that it is raining. This combines previous knowledge about the use of umbrellas with my new observation that my friend has come inside carrying an umbrella.

- *Measuring* is a way to describe something with numbers and units. Measurements can be standard or nonstandard, metric or English. Quantitative observations are made through measurement. We can measure dimensions such as time, weight, volume, length, and temperature.

- *Communicating* is telling what we know with words, diagrams, symbols, charts, or any other mode of sharing information.

- *Classifying* is organizing objects or events into groups or by rank order. Scientists classify plants and animals, rocks, galaxies, and much more.

- *Predicting* is thinking of a probable outcome based on previous observations. If I notice that the shape and size of a shadow changes throughout the day, I can predict what the shadow will look like in the future.

INTEGRATED SKILLS

- *Controlling variables* in an experiment is the practice of changing only one condition while keeping most others constant. If I want to find out what factor has the greatest impact on the rate of a pendulum's swing, I will test one variable at a time and keep the others constant. For example, I could keep the mass of the pendulum and the release angle constant, and change the length of the string.

- *Defining operationally* is determining how the object or event being observed will be measured. In the pendulum example above, the rate of the pendulum's swing could be defined operationally as the number of complete back-and-forth swings in 30 seconds.

- *Interpreting data* results in making claims based on the evidence collected.

- *Formulating a hypothesis* is making a statement about the expected outcome of an experiment. A hypothesis includes both the independent and dependent variables in the statement. The hypothesis states how changing the independent variable will affect the dependent variable. In the pendulum example a possible hypothesis would be, "Increasing the length of the string (independent variable) will decrease the rate of the pendulum's swing (dependent variable)."

- *Experimenting* is a process that begins with a testable question and ends with communicating the results of the experiment. Experiments, even when they are entirely student led, always involve a testable question, the collection of data, drawing evidence-based conclusions, and communicating the results.

- *Formulating models* is creating an evidence-based physical or mental representation of a process or concept. The model clearly and accurately represents the process or concept, can be applied under a wide variety of situations, and has predictive powers. Creating a diagram of the water cycle after collecting evidence about evaporation, condensation, transpiration, infiltration and precipitation is formulating a model. The diagram is a mental model. A model of the water cycle in which a closed jar with water in the bottom is placed on a sunny windowsill to demonstrate evaporation and condensation is a physical model. Each model clearly and accurately represents all or part of the water cycle, can be applied to a wide variety of situations, and can be used to predict what will happen. Students can use the model to explain the water cycle under a wide variety of conditions.

Appendix 1

LITERACY STRATEGIES AND TECHNIQUES

Anticipation Guide

An anticipation guide (Herber 1978) is a set of questions that are used before and after reading. Before reading, the guide activates prior knowledge, generates interest in the topic, and sets a purpose for reading (Barton and Jordan 2001). After reading, the guides can be used to assess students' comprehension of the text and concept development. The use of revised extended anticipation guides (Duffelmeyer and Baum 1992), which require students to explain the textual evidence that either supports or refutes their original choices, promotes deep engagement with the text in question. Requiring students to provide evidence that is contrary to their original thinking is one method of addressing student misconceptions.

Directed Viewing-Thinking Activity

The directed viewing-thinking activity, or DVTA, is a spin-off of the directed listening-thinking activity (DLTA), a research-based strategy for promoting comprehension of a text (Stauffer 1975). The DLTA strategy prompts students to make predictions about a text and then confirm or refute those predictions by listening to a read-aloud. We've created the DVTA to promote the development of science content knowledge by making predictions about an image, then using video clips to revise those predictions. In this way, the DVTA develops an underutilized literacy skill: the ability to critically view a diagram or video clip and obtain evidence from it.

Graphic Organizers

Graphic organizers take many forms. Commonly used graphic organizers include webs, maps, tables, and diagrams. They provide a visual representation of concepts and illustrate how concepts link together (Barton and Jordan 2001). Graphic organizers can be used in a variety of settings. When used before reading, they help students link new information to their existing schema (Ausubel 1960). As a during-reading tool, they assist students in note-taking and organizing information. Post-reading, they can be used to summarize and synthesize information or serve as a form of assessment. Novak (1991) found that when students constructed concept maps, their work better reflected their understanding of science concepts than traditional forms of testing.

Idea Circle

An idea circle is a collaborative group activity in which students pool information from multiple texts to gain conceptual understanding (Guthrie and McCann 1996). Students select books on a given topic and, within a group, no text is repeated. After reading, students share what they've learned and generate a communal store of knowledge. Idea circles are motivating for students because they allow for choice in reading material. They are also ideal for differentiated

Appendix 1

instruction in classes with varied reading levels. They are an engaging alternative to individual research and can help students understand the need to verify information with multiple sources.

Identifying Similarities and Differences

Identifying similarities and differences is one of nine research-based strategies for increasing student achievement as described by Marzano, Pickering, and Pollock (2001). This strategy enhances students' understanding of concepts. It can be achieved through comparison, classification, creating metaphors, and creating analogies (Marzano, Pickering, and Pollock 2001). Graphic organizers can be extremely useful in applying this strategy.

Making Connections to Text

Making connections is a reading comprehension strategy in which readers draw on their prior knowledge and experiences to relate to a text (Keene and Zimmerman 1997). When a reader makes connections to a text, he or she is engaged in the reading process and is actively thinking about what he or she is reading. Making connections can thus greatly enhance comprehension. Connections are classified into three categories: text-to-self, text-to-text, and text-to-world. Text-to-self connections are those that link the reader's personal experiences to what is being read. Text-to-text connections are those that link information from the current text to one that was previously read. Text-to-world connections link the text to phenomena that occur in the world at large but are not ones that the reader has necessarily experienced personally.

Mentor Texts

Mentor texts are books selected to serve as models of well-written, well-designed text (Kristo and Bamford 2004). A teacher might introduce a mentor text to an entire class within the context of a lesson, or with a small group or individual student in a writing workshop or writing conference, drawing attention to the way the author organizes information, incorporates visual elements, or uses any number of text features effectively. Students might also study texts individually as they attempt to use a feature in their own writing. Research supports the use of published texts as coaching tools. Duke and Bennett-Armistead (2003) found that if students are not exposed to much informational text, they develop informational writing skills more slowly than students who read more of the genre.

RAFT

The RAFT strategy (Santa 1988) assists students in deepening their understanding of a topic through a focused writing assignment. Santa notes that when teachers compose content-area writing prompts, they often are too broad in scope; when students respond to these prompts, they do not explain clearly or completely. RAFT prompts, composed of four key ingredients (Role of the writer, Audience, Format, and Topic), alleviate both of these problems. Almost all RAFT prompts are written by a role other than the student, to an audience other than the teacher, and in a format other than the standard essay (Santa 1988).

We've found that the specificity found in these prompts support students as they communicate their understanding of a new topic. The prompts also require students to synthesize information, rather than just writing isolated facts (Barton and Jordan 2001). We've also found that using boldface type to present the elements of a RAFT helps students focus on the necessary components in their writing.

Read-Alouds

Teacher read-alouds can be used for a variety of purposes, from introducing a text to the class as a whole to modeling fluent reading. Additionally, students of all ages simply enjoy being read to. A study of middle school students (Ivey and Broaddus 2001) found that the second favorite in-class activity was read-aloud time. Research also indicates that if students hear books read aloud, they are more likely to select them for independent reading (Martinez et al. 1997). Reading aloud from a diversity of genres can thus help promote a balanced reading diet for students. Read-alouds can be teacher directed, in which the teacher drives the conversation around the text and asks questions to which the students respond. The teacher evaluates students' responses and continues to ask questions. An alternative approach is more collaborative, in which students are free to offer their own observations, ideas, and questions. Both types of read-alouds have their place in instruction, but teachers should carefully consider the purpose of the read-aloud as they determine which approach to use.

Reader's Theater

In this oral activity, students perform a play or excerpt from a text. Each student assumes the role of a character or narrator and practices his or her part through repeated oral reading. There is no need for props, scenery, or memorization. The goal of Reader's Theater is to improve fluency and comprehension by allowing students to practice reading with expression, and giving them the opportunity to read and reread for meaning, including time to focus on word meanings (Rasinski n.d.)

Repeated Reading

As its name suggests, repeated reading is a strategy in which students reread a text or portion of a text several times. While repeated reading is most often used as a way to build fluency, it is also beneficial in terms of comprehension. With each successive reading, students are better able to comprehend as less attention is required for decoding. Teachers can also enhance comprehension by setting a new purpose for reading upon each successive rereading of the text.

Seed Discussions

A Seed Discussion (Messmer 2009) is a strategy in which students respond with writing and illustrations to a predetermined set of prompts as they are reading. The students' responses to the

Appendix 1

prompts serve as "seeds" for the discussion. During the discussion, group members share their seeds. The remaining group members comment on the seed before the next group member shares. Group members may be assigned roles to help keep the group discussions orderly and productive.

Shared Writing

Shared writing is an activity in which the teacher and students (either a small group or whole class) compose a text together, sharing their ideas and thoughts throughout the process. The teacher transcribes the text as it is orally composed. Shared writing allows students to focus on the writing process and the ideas that they would like to share (Routman 1994) and helps make the thought processes behind writing visible to students. It can be used to teach writing in various formats and to call attention to punctuation and other conventions.

REFERENCES

Ausubel, D. P. 1960. The use of advanced organizers in the learning and retention of meaningful behavior. *Journal of Educational Psychology* 51 (5): 267–272.

Barton, M. L., and D. L. Jordan. 2001. *Teaching reading in science (a supplement to the second edition of "Teaching reading in the content areas teacher's manual."* Aurora, CO: Mid-continent Research for Education and Learning.

Duffelmeyer, F. A., and D. D. Baum. 1992. Open to suggestion: The extended anticipation guide revisited. *Journal of Reading* 35 (8): 654–656.

Duke, N. K., and S. V. Bennett-Armistead. 2003. *Reading and writing informational text in the primary grades.* New York: Scholastic.

Guthrie, J. T., and A. D. McCann. 1996. Idea circles: Peer collaborations for conceptual learning. In *Lively discussions!* eds. L. Gambrell and J. Almasi, 1–13. Newark, DE: International Reading Association.

Herber, H. 1978. *Teaching reading in content areas,* 2nd ed. Englewood Cliffs, NJ: Prentice Hall.

Ivey, G., and K. Broaddus. 2001. Just plain reading: A survey of what makes students want to read in middle school classrooms. *Reading Research Quarterly* 36 (4): 350–371.

Keene, E. O., and S. Zimmerman. 1997. *Mosaic of thought: Teaching comprehension in a reader's workshop.* Portsmouth, NH: Heinemann.

Kristo, J. V., and R. A. Bamford. 2004. *Nonfiction in focus: A comprehensive framework for helping students become independent readers and writers of nonfiction, K–6.* New York: Scholastic.

Martinez, M., N. L. Roser, J. Worthy, S. Strecker, and P. Gough. 1997. Classroom libraries and children's book selections: Redefining "access" in self-selected reading. In *Inquiries in literacy theory and practice/Forty-sixth yearbook of the National Reading Conference,* eds. C. K. Kinzer, K. A. Hinchman, and D. J. Leu, 265–272. Chicago: National Reading Conference.

Marzano, R. J., D. J. Pickering, and J. E. Pollock. 2001. *Classroom instruction that works: Research-based strategies for increasing student achievement.* Alexandria, VA: Association for Supervision and Curriculum Development.

Appendix 1

Messmer, A. M. 2009. Seed Discussions. *http://literacy.purduecal.edu/STUDENT/ammessme/Seed.html*

Novak, J. D. 1991. Clarifying with concept maps. *The Science Teacher* 58 (7): 45–49.

Padilla, M. J. 1990. The science process skills. *Research Matters—to the Science Teacher,* No. 9004. *www.narst.org/publications/research/skill.cfm*

Rasinski, T. n.d. *Reader's Theater. www.literacyconnections.com/rasinski-readers-theater.php*

Rezba, R. J., C. Sprague, and R. L. Fiel. 2003. *Learning and assessing science process skills.* Dubuque, IA: Kendall/Hunt.

Routman, R. 1994. *Invitations: Changing as teachers and learners K–12.* Portsmouth, NH: Heinemann.

Santa, C. M. 1988. *Content reading including study systems: Reading, writing and studying across the curriculum*. Dubuque, IA: Kendall/Hunt. *www.eric.ed.gov/PDFS/ED372363.pdf*

Stauffer, R. G. 1975. *Directing the reading-thinking process*. New York: HarperCollins.

Rubrics and Achievement Grading Standards

SCIENCE AND LITERACY RUBRIC

Criterion	Criterion Exceeded	Criterion Met	Criterion Not Sufficiently Met
Science Content	Illustrates an accurate and thorough understanding of scientific concepts as defined by the objectives *and* is able to tie concepts together in a coherent way.	Illustrates an accurate and thorough understanding of scientific concepts as defined by the objectives.	Illustrates a limited or inaccurate understanding of scientific concepts as defined by the objectives.
Scientific Terminology (vocabulary)	Demonstrates correct, developmentally appropriate understanding of all terminology through writing and/or speaking *and* is able to draw relationships between the terms.	Demonstrates correct, developmentally appropriate understanding of all terminology through writing and/or speaking.	Demonstrates limited or inaccurate understanding of scientific terms.
Use of Evidence (from texts and investigations)	Explicitly refers to evidence in work products (e.g., RAFT prompt, explanatory report).	Consistently alludes to evidence in work products.	Work products are not supported by evidence.
Clarity of Message	Writing, using visual images, and/or speaking is coherent, complete, and conveys sufficient information to elaborate on the points raised.	Writing, using visual images, and/or speaking is coherent, complete, and conveys sufficient information to make the point.	Writing, using visual images, and/or speaking lacks coherence, is incomplete, and/or includes insufficient information.
Organization	The organization of the work product enhances the meaning conveyed in it.	The components of the work product are presented in a logical sequence and arrangement.	The components of the work product are not presented in a logical sequence and arrangement.
Format	The work product adheres to the assigned format *and* includes additional features that enhance the product.	The work product adheres to the assigned format.	The work product does not adhere to the assigned format.
Quality of Work	The work product exceeds the student's normal standard of work.	The work product meets the student's normal standard of work.	The work product falls below the student's normal standard of work.

Appendix 2

ACHIEVEMENT GRADING STANDARDS

Percent (grade)	Standard
100%	Based on the rubric, the student has exceeded all of the criteria.
95%	Based on the rubric, the student has met all and exceeded many of the criteria.
90%	Based on the rubric, the student has met all and exceeded some of the criteria.
85%	Based on the rubric, the student has met all of the criteria.
80%	Based on the rubric, the student has met most of the criteria.
75%	Based on the rubric, the student has met some of the criteria.

SCIENCE PROCESS SKILLS RUBRIC

	Independent	Developing	Beginning
Basic Science Process Skills			
Observing	The student independently makes and records detailed observations using his or her senses and appropriate measuring tools. Recorded observations are clear and free of ambiguous language and illustrations.	With teacher or peer assistance, the student makes and records detailed observations using his or her senses and appropriate measuring tools. Recorded observations are described in clear language and illustrations.	With teacher assistance, the student makes and records simple observations using his or her senses and appropriate measuring tools. Recorded observations may include ambiguous language and illustrations.
Measuring	Working independently with appropriate tools, the student makes and records detailed measurements.	Using appropriate tools and with teacher or peer assistance, the student makes and records detailed measurements.	Using appropriate tools and with teacher assistance, the student makes and records simple measurements.
Classifying	Working independently, the student sorts and organizes objects by group or order, and documents the process in words and with illustrations/diagrams/flow charts.	With teacher or peer assistance, the student sorts and organizes objects by group or order and describes the process in words and/or illustrations/diagrams/flow charts.	With teacher assistance, the student sorts and organizes objects by group or order and describes the process in words.
Inferring	Working independently the student clearly explains a possible relationship between previous knowledge and new observations.	With teacher or peer assistance, the student explains a possible relationship between previous knowledge and new observations.	With teacher assistance, the student explains a possible relationship between previous knowledge and new observations.

Appendix 2

	Independent	Developing	Beginning
Predicting	Working independently, the student makes well-reasoned predictions based on previous observations.	With teacher or peer assistance, the student makes thoughtful predictions based on previous observations.	With teacher assistance, the student makes predictions based on previous observations.
Communicating	Working independently, the student communicates detailed information with words and/or diagrams in a format of his or her choosing.	With teacher or peer assistance, the student communicates basic information with words and/or simple diagrams in a format of his or her choosing.	With teacher assistance, the student communicates basic information with words and/or simple diagrams in a prescribed format.
Integrated Science Process Skills			
Controlling variables	Working independently, the student identifies dependent and independent variables.	With teacher or peer assistance, the student identifies dependent and independent variables.	With teacher assistance, the student identifies dependent and independent variables.
Defining operationally	Working independently, the student determines how the dependent variable will be measured.	With teacher or peer assistance, the student determines how the dependent variable will be measured.	With teacher assistance, the student determines how the dependent variable will be measured.
Interpreting data	Working independently, the student makes evidence-based claims.	With teacher or peer assistance, the student makes evidence-based claims.	With teacher assistance, the student makes evidence-based claims.
Formulating hypotheses	Working independently, the student makes a cause-and-effect statement about the expected outcome of the experiment.	With teacher or peer assistance, the student makes a cause-and-effect statement about the expected outcome of the experiment.	With teacher assistance, the student makes a cause-and-effect statement about the expected outcome of the experiment.
Experimenting	Working independently, the student designs and conducts an experiment.	With teacher or peer assistance, the student designs and conducts an experiment.	With teacher assistance, the student conducts a teacher-led experiment.
Formulating models	Working independently, the student creates an evidence-based representation of a process or concept.	With teacher or peer assistance, the student creates an evidence-based representation of a process or concept.	With teacher assistance, the student creates an evidence-based representation of a process or concept.

Name_____ Date_____

MY SCIENCE PROCESS SKILLS RUBRIC

Place a check mark beside the statement that tells how you feel about your ability to do each of the following things.

Observing: I can make and record detailed observations using my senses and appropriate measuring tools. My observations are clearly described with words and illustrations/diagrams.		I need my teacher's help.
		I need my classmate's help.
		I can do this by myself.
I can strengthen my skills by		

Measuring: Using appropriate tools, I can make and record detailed measurements.		I need my teacher's help.
		I need my classmate's help.
		I can do this by myself.
I can strengthen my skills by		

Classifying: I can sort and organize objects by group or order, and document the process in words and with illustrations/diagrams/flow charts.		I need my teacher's help.
		I need my classmate's help.
		I can do this by myself.
I can strengthen my skills by		

Inferring: I can clearly explain a possible relationship between what I already know and new observations.		I need my teacher's help.
		I need my classmate's help.
		I can do this by myself.
I can strengthen my skills by		

Predicting: I can make reasonable predictions based on previous observations.		I need my teacher's help.
		I need my classmate's help.
		I can do this by myself.
I can strengthen my skills by		

Appendix 2

Name_____ Date_____

Communicating: I can share detailed information with words and/or illustrations/diagrams in a format that I pick.		I need my teacher's help.
		I need my classmate's help.
		I can do this by myself.
I can strengthen my skills by		

Controlling variables: I can identify dependent and independent variables.		I need my teacher's help.
		I need my classmate's help.
		I can do this by myself.
I can strengthen my skills by		

Defining operationally: I can describe how the dependent variable will be measured.		I need my teacher's help.
		I need my classmate's help.
		I can do this by myself.
I can strengthen my skills by		

Interpreting data: I can make evidence-based claims.		I need my teacher's help.
		I need my classmate's help.
		I can do this by myself.
I can strengthen my skills by		

Making hypotheses: I can make a cause-and-effect statement about the expected outcome of the experiment.		I need my teacher's help.
		I need my classmate's help.
		I can do this by myself.
I can strengthen my skills by		

Experimenting: I can design and conduct an experiment.		I need my teacher's help.
		I need my classmate's help.
		I can do this by myself.

NATIONAL SCIENCE TEACHERS ASSOCIATION

Name_____ Date_____

I can strengthen my skills by		
Making models: I can make an evidence-based model of a process or concept.		I need my teacher's help.
		I need my classmate's help.
		I can do this by myself.
I can strengthen my skills by		

Standards Alignment Matrices

NATIONAL SCIENCE EDUCATION STANDARDS

Standard	Scientists Like Me	Measuring Pennies and More	Minds-on Matter	Classroom Curling	Beaks and Biomes	My Favorite Tree	Come Fly With Me	Drip Drop Detectives	Let's Dig!	Patterns in the Sky
K–4 Science as Inquiry										
Abilities Necessary to Do Scientific Inquiry										
Ask a question about objects, organisms, and events in the environment.	✓	✓	✓		✓	✓	✓		✓	
Plan and conduct a simple investigation.	✓			✓			✓		✓	
Employ simple equipment and tools to gather data and extend the senses.		✓	✓	✓	✓			✓	✓	
Use data to construct a reasonable explanation.	✓	✓	✓		✓	✓	✓	✓	✓	✓
Communicate investigations and explanations.	✓	✓	✓		✓	✓	✓	✓	✓	
Understanding About Scientific Inquiry										
Scientific investigations involve asking and answering a question and comparing the answer with what scientists already know about the world.				✓						✓
Simple instruments, such as magnifiers, thermometers, and rulers, provide more information than scientists obtain using only their senses.				✓						
Scientists develop explanations using observations (evidence) and what they already know about the world (scientific knowledge). Good explanations are based on evidence from investigations.										✓
5–8 Science as Inquiry										
Abilities Necessary to Do Scientific Inquiry										
Identify questions that can be answered through scientific investigations.	✓			✓	✓	✓				
Design and conduct a scientific investigation.	✓			✓	✓	✓	✓	✓	✓	
Use appropriate tools and techniques to gather, analyze, and interpret data.	✓			✓	✓	✓	✓	✓	✓	

	Scientists Like Me	Measuring Pennies and More	Minds-on Matter	Classroom Curling	Beaks and Biomes	My Favorite Tree	Come Fly With Me	Drip Drop Detectives	Let's Dig!	Patterns in the Sky
Develop descriptions, explanations, predictions, and models using evidence.	✓	✓	✓	✓	✓		✓		✓	
Think critically and logically to make the relationships between evidence and explanations.	✓	✓	✓	✓	✓	✓	✓	✓	✓	
Recognize and analyze alternative explanations and predictions.	✓									
Communicate scientific procedures and explanations.	✓							✓		
Use mathematics in all aspects of scientific inquiry.										
K–4 History and Nature of Science *Science as a Human Endeavor* Science and technology have been practiced by people for a long time.	✓									
Men and women have made a variety of contributions throughout the history of science and technology.	✓									
Although men and women using scientific inquiry have learned much about the objects, events, and phenomena in nature, much more remains to be understood. Science will never be finished.	✓									
Many people choose science as a career and devote their entire lives to studying it. Many people derive great pleasure from doing science.	✓									
5–8 History and Nature of Science *Science as a Human Endeavor* Women and men of various social and ethnic backgrounds—and with diverse interests, talents, qualities, and motivations—engage in the activities of science, engineering, and related fields such as the health professions. Some scientists work in teams, and some work alone, but all communicate extensively with others.	✓									
Science requires different abilities, depending on such factors as the field of study and type of inquiry. Science is very much a human endeavor, and the work of science relies on basic human qualities, such as reasoning, insight, energy, skill, and creativity—as well as on scientific habits of mind, such as intellectual honesty, tolerance of ambiguity, skepticism, and openness to new ideas.	✓									

Appendix 3

	Scientists Like Me	Measuring Pennies and More	Minds-on Matter	Classroom Curling	Beaks and Biomes	My Favorite Tree	Come Fly With Me	Drip Drop Detectives	Let's Dig!	Patterns in the Sky
History of Science Many individuals have contributed to the traditions of science. Studying some of these individuals provides further understanding of scientific inquiry, science as a human endeavor, the nature of science, and the relationships between science and society.	✓									
In historical perspective, science has been practiced by different individuals in different cultures. In looking at the history of many peoples, one finds that scientists and engineers of high achievement are considered to be among the most valued contributors to their culture.	✓									
Tracing the history of science can show how difficult it was for scientific innovators to break through the accepted ideas of their time to reach the conclusions that we currently take for granted.	✓									
K–4 Physical Science *Properties of Objects and Materials* Objects have many observable properties, including size, weight, shape, color, temperature, and the ability to react with other substances. Those properties can be measured using tools, such as rulers, balances, and thermometers.			✓							
Materials can exist in different states—solid, liquid, and gas. Some common materials, such as water, can be changed from one state to another by heating or cooling.			✓							
Position and Motion of Objects The position of an object can be described by locating it relative to another object or the background.				✓						
An object's motion can be described by tracing and measuring its position over time.				✓						
The position and motion of objects can be changed by pushing or pulling. The size of the change is related to the strength of the push or pull.				✓						

Standard	Scientists Like Me	Measuring Pennies and More	Minds-on Matter	Classroom Curling	Beaks and Biomes	My Favorite Tree	Come Fly With Me	Drip Drop Detectives	Let's Dig!	Patterns in the Sky
5–8 Physical Science — *Properties and Changes of Properties in Matter* — A substance has characteristic properties, such as density, a boiling point, and solubility, all of which are independent of the amount of the sample.			✓							
Motions and Forces — An object that is not being subjected to a force will continue to move at a constant speed and in a straight line.				✓						
If more than one force acts on an object along a straight line, then the forces will reinforce or cancel one another, depending on their direction and magnitude. Unbalanced forces will cause changes in the speed or direction of an object's motion.				✓						
K–4 Life Science — *Characteristics of Organisms* — Organisms have basic needs. For example, animals need air, water, and food; plants require air, water, nutrients, and light. Organisms can survive only in environments in which their needs can be met. The world has many different environments, and distinct environments support the life of different types of organisms.					✓					
Each plant or animal has different structures that serve different functions in growth, survival, and reproduction. For example, humans have distinct body structures for walking, holding, seeing, and talking.					✓	✓	✓			
5–8 Life Science — *Diversity and Adaptations of Organisms* — Biological evolution accounts for the diversity of species developed through gradual processes over many generations. Species acquire many of their unique characteristics through biological adaptation, which involves the selection of naturally occurring variations in populations. Biological adaptations include changes in structures, behaviors, or physiology that enhance survival and reproductive success in a particular environment.					✓					

	Scientists Like Me	Measuring Pennies and More	Minds-on Matter	Classroom Curling	Beaks and Biomes	My Favorite Tree	Come Fly With Me	Drip Drop Detectives	Let's Dig!	Patterns in the Sky
Millions of species of animals, plants, and microorganisms are alive today. Although different species might look dissimilar, the unity among organisms becomes apparent from an analysis of internal structures, the similarity of their chemical processes, and the evidence of common ancestry.						✓				
Extinction of a species occurs when the environment changes and the adaptive characteristics of a species are insufficient to allow its survival. Fossils indicate that many organisms that lived long ago are extinct. Extinction of species is common; most of the species that have lived on the Earth no longer exist.									✓	
Structure and Function in Living Systems Living systems at all levels of organization demonstrate the complementary nature of structure and function. Important levels of organization for structure and function include cells, organs, tissues, organ systems, whole organisms, and ecosystems.							✓			
K–4 Earth and Space Science *Properties of Earth Materials* Fossils provide evidence about the plants and animals that lived long ago and the nature of the environment at that time.									✓	
Objects in the Sky The Sun, Moon, stars, clouds, birds, and airplanes all have properties, locations, and movements that can be observed and described.										✓
5–8 Earth and Space Science *Structure of the Earth System* Water, which covers the majority of the Earth's surface, circulates through the crust, oceans, and atmosphere in what is known as the "water cycle." Water evaporates from the Earth's surface, rises and cools as it moves to higher elevations, condenses as rain or snow, and falls to the surface where it collects in lakes, oceans, soil, and in rocks underground.								✓		
Earth's History Fossils provide important evidence of how life and environmental conditions have changed.									✓	

	Scientists Like Me	Measuring Pennies and More	Minds-on Matter	Classroom Curling	Beaks and Biomes	My Favorite Tree	Come Fly With Me	Drip Drop Detectives	Let's Dig!	Patterns in the Sky
K–4 Science and Technology *Abilities of Technological Design* Propose a solution.			✓							
Implement proposed solutions.			✓							
Evaluate a product or design.			✓							
Communicate a problem, design, and solution.			✓							
5–8 Science and Technology *Abilities of Technological Design* Design a solution or product.			✓							
Implement a proposed design.			✓							
Evaluate completed technological design or products.			✓							
Communicate the processes of technological design.			✓							
K–4 Science in Personal and Social Perspectives *Types of Resources* The supply of many resources is limited. If used, resources can be extended through recycling and decreased use.								✓		

Source: National Research Council (NRC). 1996. *National science education standards.* Washington, DC: National Academies Press.

Appendix 3

COMMON CORE STATE STANDARDS FOR ENGLISH LANGUAGE ARTS

Standard	Patterns in the Sky	Let's Dig!	Drip Drop Detectives	Come Fly With Me	My Favorite Tree	Beaks and Biomes	Classroom Curling	Minds-on Matter	Measuring Pennies and More	Scientists Like Me
Informational Text										
Key Ideas and Details										
Grade 3: Ask and answer questions to demonstrate understanding of a text, referring explicitly to the text as the basis for the answers.			✓							✓
Grade 3: Describe the relationship between a series of historical events, scientific ideas or concepts, or steps in technical procedures in a text, using language that pertains to time, sequence, and cause/effect.							✓	✓		
Grade 4: Refer to details and examples in a text when explaining what the text says explicitly and when drawing inferences from the text.										✓
Grade 4: Explain events, procedures, ideas, or concepts in a historical, scientific, or technical text, including what happened and why, based on specific information in the text.			✓				✓	✓		
Grade 5: Quote accurately from a text when explaining what the text says explicitly and when drawing inferences from the text.										✓
Grade 5: Explain the relationships or interactions between two or more individuals, events, ideas, or concepts in a historical, scientific, or technical text based on specific information in the text.								✓		
Craft and Structure										
Grade 3: Determine the meaning of general academic and domain-specific words and phrases in a text relevant to a grade 3 topic or subject area.			✓	✓			✓			
Grade 3: Use text features and search tools (e.g., key words, sidebars, hyperlinks) to locate information relevant to a given topic efficiently.							✓			
Grade 4: Determine the meaning of general academic and domain-specific words or phrases in a text relevant to a grade 4 topic or subject area.			✓	✓			✓			

Standard	Patterns in the Sky	Let's Dig!	Drip Drop Detectives	Come Fly With Me	My Favorite Tree	Beaks and Biomes	Classroom Curling	Minds-on Matter	Measuring Pennies and More	Scientists Like Me
Grade 4: Compare and contrast a firsthand and secondhand account of the same event or topic; describe the differences in focus and the information provided.			✓							
Grade 5: Determine the meaning of general academic and domain-specific words and phrases in a text relevant to a grade 5 topic or subject area.			✓	✓	✓		✓			
Grade 5: Analyze multiple accounts of the same event or topic, noting important similarities and differences in the point of view they represent.			✓							
Integration of Knowledge and Ideas Grade 3: Describe the logical connection between particular sentences and paragraphs in a text (e.g., comparison, cause/effect, first/second/third in a sequence).								✓		
Grade 3: Use information gained from illustrations (e.g., maps, photographs) and the words in a text to demonstrate understanding of the text (e.g., where, when, why, and how key events occur).	✓		✓	✓						
Grade 4: Integrate information from two texts on the same topic in order to write or speak about the subject knowledgeably.			✓							
Grade 4: Interpret information presented visually, orally, or quantitatively (e.g., in charts, graphs, diagrams, time lines, animations, or interactive elements on web pages) and explain how the information contributes to an understanding of the text in which it appears.	✓									
Grade 5: Integrate information from several texts on the same topic in order to write or speak about the subject knowledgeably.			✓						✓	✓
Writing *Text Types and Purposes* Grades 3–5: Write informative/explanatory texts to examine a topic and convey ideas and information clearly.		✓	✓	✓	✓	✓	✓			✓

Standard	Patterns in the Sky	Let's Dig!	Drip Drop Detectives	Come Fly With Me	My Favorite Tree	Beaks and Biomes	Classroom Curling	Minds-on Matter	Measuring Pennies and More	Scientists Like Me
Grade 3–5: Write narratives to develop real or imagined experiences or events using effective technique, descriptive details, and clear event sequences.	✓									
Production and Distribution of Writing Grades 3–5: With guidance and support from adults, produce writing in which the development and organization are appropriate to task and purpose.		✓				✓				
Grades 3–5: With guidance and support from peers and adults, develop and strengthen writing as needed by planning, revising, and editing.		✓				✓				
Research to Build and Present Knowledge Grades 3–5: Conduct short research projects that build knowledge about a topic.		✓	✓			✓				
Grades 3–5: Recall information from experiences or gather information from print and digital sources; take brief notes on sources and sort evidence into provided categories.		✓	✓			✓				
Grades 4–5: Draw evidence from literary or informational texts to support analysis, reflection, and research.		✓				✓				
Speaking and Listening *Comprehension and Collaboration* Grades 3–5: Engage effectively in a range of collaborative discussions (one-on-one, in groups, and teacher-led) with diverse partners on grade-appropriate topics and texts, building on others' ideas and expressing their own clearly.	✓			✓						✓

Source: National Governors Association (NGA) Center for Best Practices, Council of Chief State School Officers (CCSSO). 2010. *Common core state standards English language arts.* Washington DC: NGA Center for Best Practices, CCSSO.

Index

*Page numbers printed in **boldface** type refer to tables and figures.*

Index

Index

Index

Index

Forces and motion. *See* "Classroom Curling"

Formative assessment, 15, 17, 36–37, 53, 57, 58, 74, 78, 99, 102, 122, 128, 130, 148, 152, 153, 163, 165, 166, 180, 183, 184, 186, 199, 204, 206, 229, 231, 233, 249, 255, 258, 260

Formative assessment probes, 17
   "Darkness at Night," 253, 255, 256, 265
   "Me and My Shadow," 251, 252, 255, 256, 264

Formulating a hypothesis, 270, 278

Formulating models, 270, 278

The Fossil Feud: Marsh and Cope's Bone Wars, **41,** 223, 233

Fossils. *See* "Let's Dig!"

Fossils (Stewart), **41,** 223

Fossils (Walker), **41,** 223

Fossils Tell of Long Ago, **41,** 223

Foucault, Jean-Bernard-Léon, 45, 47, 49, 56–57

Freezing. *See* "Minds-on Matter"

Friction, 26, **39,** 111, 112–113, 115, 126, 127, 128, 129, 130, 136, 139. *See also* "Classroom Curling"

Fries-Gaither, Jessica, viii, 1, 25

Frogs, 168

Frogs and Toads, 168

Frogs Sing Songs, 168

The Frozen Tundra, 149

Fujita-Pearson Tornado Intensity Scale, 66, 70

G

Galileo, 242, 246, 249, 260

Gases. *See* "Minds-on Matter"

Gerlovich, J., 15

Gibbons, Gail, 95

Goldish, Meish, 224

Grading: "Achievement Grading Standards," 37, 58, 59, 80, 104, 130, 131, 153, 166, 186, 187, 205, 233, 256, 258, 260, 277

Graphic organizers, 2, 14, 271, 272
   "Characteristics of My Tree," 162, 164, 165, 170
   "Defining Friction," 118, 127, 128, 136

"Defining Gravity," 118, 128, 137

DVTA, 179, 184, 190

"Idea Circle," 145, 149, 152, 155

"Newton and Me Connections," 118, 129, 130, 138

"Physical Properties," 97, 99, 109

"Seed Discussion," 71, 77, 78–79, 85

"Site Report," 226, 232, 233, 238–239

types of, 272

Venn diagram, 21, 145, 152, 153, 156

"Who Is a Scientist?", 52, 53, 61

Gravity, 31, **39, 67,** 111, 112, 115, 127, 128, 129, 130, 137, 139. *See also* "Classroom Curling"

Gray, Susan H., 222

Gregor Mendel: The Friar Who Grew Peas, **38,** 50, 56

H

Hands-on activities, 1
   disjointed series of, 18
   vs. inquiry-based instruction, 14
   in learning cycle, 17–18
   vs. minds-on science, 268
   timing and sequence of reading and, 8

Harbo, Christopher L., 177

Harbor, J., 193–194

Harste, J. C., 31

Harvey, Stephanie, 8

Hawkins, Benjamin Waterhouse, 222, 233

Hint cards for scavenger hunts, **38,** 71, 72, 73, **73**

Hooks, Gwendolyn, 149

How Do You Measure Length and Distance?, **38,** 69, 76

How Do You Measure Liquids?, **38,** 69, 76

How Do You Measure Time?, **38,** 69, 76

How Do You Measure Weight?, **38,** 69, 76

How Tall, How Short, How Far Away, **38,** 69, 76

How-to texts, 27, 29, **30**

Hunter, Ryan Ann, 26, 177

Hypothesis formulation, 270, 278

Index

Index

Index

Index

Index

Index

Precipitation, 192, **193,** 195–196, 203, 205, 216, 270
Predicting, 270, 278
Priddy, M., 193–194
Professional development, 1, 2, 7
Purslow, Frances, **41,** 196

Q

Questions, testable, 1, 11, 12, **13,** 14, 16, 29, 36, 45, 56, 57, 59, 78, 270

R

Rachel: The Story of Rachel Carson, **38,** 50, 55
RAFT writing prompts, 141, 145, 153, 157, 272–273, 276
Ranger Rick, 25
Ray, Deborah Kogan, 222
Read-alouds, 2, 16, 19, 22, 51, 55, 57, 130, 147, 173, 180, 186, 228, 271, 273
Reader of the Rocks, **38,** 51
Reader's theater, 256, 258, 273
Reading, 20
    concept-oriented reading instruction, 25
    in context of learning cycle, 20
    paired, 76, 77
    repeated, 45, 58, 273
    of science texts, 1
    timing and sequences of inquiry and, 9
Reading comprehension, 8, 19, 20, 45, 54, 129, 148, 152, 173, 271, 272, 273
Reading level of books, 35
Reading strategies, 9, 20, 45, 128. *See also* Literacy strategies and techniques
Red Knot: A Shorebird's Incredible Journey, **40,** 144, 153
Reference texts, 26, 29, **30,** 31
Repeated reading, 45, 58, 273
Reproducibility of results, 193
Riddle cards for scavenger hunts, **38,** 71, 72, **72,** 73

Robbins, Chandler S., 177
Role, Audience, Format, Topic (RAFT) writing prompts, 141, 145, 153, 157, 272–273, 276
Royston, Angela, 195
Rubrics, 276–281
    "My Science Process Skills Rubric," 279–281
    "Science and Literacy Rubric," 276–277
    "Science Process Skills Rubric," 277–278
Rushby, Pamela, 225
Rushworth, P., 113

S

S Is for Scientists: A Discovery Alphabet, 28, **38,** 51, 57
Sadler, Wendy, 115
Safety considerations, 52, 71, 97, 118, 146, 162, 179, 197, 226, 248
Sanderling adaptation and migration. *See* "Beaks and Biomes"
Scavenger hunts, 16, 65
    measurement, 71, **72,** 72–74, **73**
Schellenberger, L., 193–194
Schmidt, B., 37
Science, instructional time for, 2, 7
"Science and Literacy Rubric," 20, 58, 59, 80, 104, 130, 153, 166, 186, 187, 205, 233, 256, 258, 260, 276–277
Science journals, **38, 39,** 51, 96, 99, 100, 102, 103, 229, 230, 231, 233. *See also* Science notebooks
Science-literacy connection, 7–9
    language differences and, 8
    rationale for, 8–9
    timing and sequences of instructional activities, 9
Science notebooks, 14, 17, **40, 99,** 122, 124, 127, 128, 154, 162, 163. *See also* Science journals
Science process skills, 8, 9, 16, 37, **38, 40,** 45, 53–59, 141, 173, 269–270
    assessment of, 57, 59, 79, 105, 154, 261
        rubrics for, 277–281

Index

Index

Index

NATIONAL SCIENCE TEACHERS ASSOCIATION